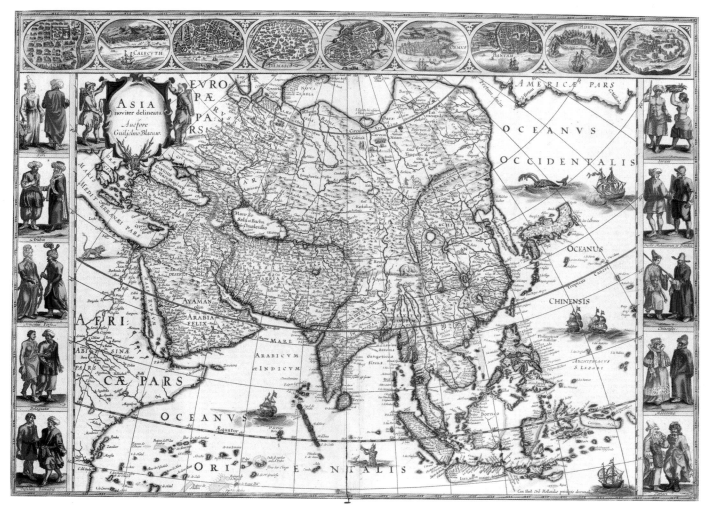

TO THE ENDS OF THE
EARTH
100 MAPS THAT CHANGED THE WORLD

JEREMY HARWOOD

WITH AN INTRODUCTION BY SARAH BENDALL

CHARTWELL
BOOKS, INC.

This edition published in 2011 by

CHARTWELL BOOKS, INC.
A division of BOOK SALES, INC.
276 Fifth Avenue Suite 206
New York, New York 10001
USA

ISBN-13: 978-0-7858-2898-3

Conceived, edited and designed by
Marshall Editions
The Old Brewery
6 Blundell Street
London N7 9BH

Publisher: Richard Green
Commissioning Editor: Claudia Martin
Art Director: Ivo Marloh
Design: Melissa Alaverdy, Phil Himmelmann
Project Editor: Johanna Geary
Picture Manager: Veneta Bullen
Indexer: Hilary Bird
Production: Anna Pauletti

Originated in Hong Kong by Modern Age
Printed and bound in Singapore by Star Standard
Industries (PTE) Ltd

The publishers and author gratefully acknowledge the
advice of the following:

Dr Sarah Bendall, Fellow of Emmanuel College, Cambridge,
editor of The Dictionary of Land Surveyors and
Mapmakers of Britain and Ireland, a contributor to Rural
Images: Estate Maps in the Old and New Worlds, and
author of Maps, Land and Society: A History;

Dr Christopher Board O.B.E., Senior Lecturer in Geography at
the London School of Economics until his retirement, and
a contributor to Sheetlines, the journal of the Charles
Close Society for the Study of Ordnance Survey Maps;

Dr Catherine Delano-Smith, editor of Imago Mundi, the Journal for
the History of Cartography, co-author of English Maps:
A History, and a senior research fellow at the Institute of
Historical Research, London;

Professor P.D.A. Harvey F.B.A., Professor Emeritus of Medieval
History at Durham University, whose recent books include
Medieval Maps and Mappa Mundi: The Hereford
World Map; and

Professor Roger Kain C.B.E. F.B.A., Montefiore Professor
of Geography at the University of Exeter, whose
publications include The Enclosure Maps of England
and Wales and English Maps: A History.

CONTENTS

INTRODUCTION

T his book is based on the premise that maps can change the world. This is a bold assertion: how do maps make the changes, and whose world is thereby altered? Maps are a particular way of representing the world. In their *History of Cartography,* the late Brian Harley and David Woodward defined them as "graphic representations that facilitate a spatial understanding of things, concepts, conditions, processes, or events in the human world." Maps are the products of actions by particular social groups, each of which exists in its own historical context. But maps do not just reflect these historical circumstances: as part of the power structures, organization, and administration of societies, they have a reciprocal relationship with them. Maps are both predictive and descriptive, they are proactive and reactive, they can push both physical and intellectual boundaries, and they can encourage research and debate. This is as true for maps made in prehistoric times as for those produced today. For example, maps could reinforce, influence, or alter contemporary beliefs. Thus the first map illustrated in this book, the Babylonian World Map (Map 1), shows the world (with Babylon at its center) and its encircling earthly ocean as believed by the Mesopotamians in c. 600 B.C.E.; the *mappae mundi* in Chapter 3, "The Medieval World," show medieval concepts of the world and were used to convey much theological (see, for example, Map 16), moral, historical, and zoological (Map 17) information in a spatial framework; while the Peter's projection (Map 96) gave 20th-century map-users much to think about when seeing new shapes and relative sizes of continents.

Above: A 16th-century woodcut depicts the use of surveying techniques.

So how can maps change the world? At the global scale, maps have been used to convey information about new lands, trade routes, and colonial territories, for instance the Cantino Planisphere (Map 31) or Cook's map of New Zealand (Map 64). Thus worlds both expanded—by including discoveries of new lands—and contracted, by making distant parts seem nearer and easier to govern. While it cannot be claimed that maps alone formed overseas empires, they played important parts in helping colonization and in convincing those at home of the legitimacy and enforceability of their imperial claims. For example, the 16th-century Cantino Planisphere gives prominence to the lands claimed by the Portuguese, for whom the information was collected, and indeed claims Newfoundland and Greenland for them. Later, in the 17th century, the Dutch and English were in conflict with each other over territorial claims in North America. Maps were used to express the ideological agendas of both sides and each avoided using the other's place-names. John Smith's

Opposite: A wallpainting dating from 1413–1403 B.C.E., from the tomb of Mennah, scribe under Pharaoh Thutmose IV, depicts a surveyor measuring a cornfield in the presence of the owner.

map of New England (Map 42) showed the English view of the area. The local native inhabitants should not be forgotten: by aiding colonial expansion, maps affected the worlds of the indigenous population as well as those of the people who came to live in the newly claimed territories.

Ever since the Renaissance, collections of maps bound together in volumes have both reflected and influenced contemporary knowledge, such as Ptolemy's maps (Map 7), Mercator's *Atlas* (Map 29), Ortelius's *Theatrum Orbis Terrarum* (Map 43), or Waghenaer's sea atlas (Map 46). These atlases could perpetuate misperceptions of countries for many years after they were made: the Ptolemaic vision of the world dominated cartographic and geographic thought for centuries, while Raleigh's legendary city of El Dorado (Map 38) acquired its location in Guiana through misinterpretation by Europeans of information from indigenous peoples and continued to be shown on Western maps until the 1840s.

Below: This 19th-century brass sextant, with its tripod stand and telescopic sights, could be used to find latitude by measuring the height of the midday sun or of a suitable star at night.

Maps were used to help in the discovery of new worlds by aiding navigation, and maps also enabled travel in general to take place much more easily. The medieval Carte Pisane (Map 21) is an early example of a portolan chart used by navigators, while the projection known as Mercator's (Map 29) made it much easier for sailors from the 16th century onward to follow a compass course. Travel was not necessarily by explorers. The Peutinger Table (Map 6) shows routes that had been used by those traveling within the Roman Empire, the monk Matthew Paris showed pilgrim routes to the Holy Land (Map 23), and in 1675 John Ogilby devised an English road atlas showing roads as strips (Map 51), a format that still continues today. From the 20th century, Beck's map of the London underground (Map 94) is an example of a schematic map that enabled users to find their way along a complex network of routes, while the A–Z map of London (Map 93) changed the worlds of local travelers wanting to find their way in a city. Maps have also changed the world of the tourist: of the cyclist in Britain, for example (Map 82), of the driver through France (Map 84), or of the visitor to Rome (Map 83).

Surveyors used maps to lay out new lands. Maps were an integral part of the late 18th-century English enclosure process (Map 62), for plans showed how the land was to be redistributed and allocated. Christopher Wren's map of London after the fire of 1666 (Map 55) is an example of how maps can be used to inform planning issues and to show possible new layouts of cities, though in the event his scheme was not carried out. Maps helped, and still assist, in the construction of canals, railroads, and roads by, for example, showing the land through which they will pass and the obstacles that need to be overcome, and then by publicizing the resulting network of routes (Map 75). Thus maps were agents in changing the physical landscape as well as in altering the lives of those who lived in it and traveled through it.

Boundary disputes could be settled using maps. In 1768, Charles Mason and Jeremiah Dixon determined the boundary between Pennsylvania and Maryland, and the line, the "Mason-Dixon line," came to represent the division between the southern slave states and northern free states (Map 60). John Mitchell's map of North America was used by the American and British negotiators in Paris in 1782–3 to define the boundaries of the United States, and continued to be used in this capacity right up to the late 20th century (Map 59).

Maps are expressions of power, authority, and control. Papal power and ambition were displayed on the walls of the Vatican in the 16th century (Map 48) and used to impress and influence visitors. Christopher Saxton's county atlas of England and Wales (Map 50) showed the power and authority of the English crown and helped in the growth of cartographic awareness among the landowning classes in Elizabethan England. The decorative aspects of maps, making them attractive objects for display, could be used to show wealth and prestige. Blaeu's *Atlas Major*, for instance (Map 47), was available in luxurious presentation copies and was traditionally given by the United Provinces to royal personages in the 17th century.

Maps were, and still are, used as propaganda, and could exaggerate, suppress, or falsify information to create a particular impression. The late 16th-century "Christian knight map" (Map 45), for example, was used as propaganda by the Dutch against the Spanish in the Dutch struggle for independence. A 20th-century example is the Indian propaganda map drawn during World War I (Map 89). And the map of Sultan Njoya's kingdom in Cameroon (Map 73) was used to present an image of a stable state and to conceal power struggles in 1916. In warfare, maps were used to show how troops were deployed, for example in the map of the Second Battle of Bull Run of 1862 (Map 77), and to give retrospective insights into what actually happened on the ground (Map 76); they helped individuals to find their way through enemy territory, for example the silk maps used during World War II (Map 87); and they shared out land at the end of hostilities, as in the case of the partition of Berlin (Map 91).

While many of the above examples show how maps could change the worlds of whole societies and, by extension, of the individuals within them, there are other instances of how maps changed local worlds. In the late 19th century, the lives of the poor of London, for example, were changed for the better as a result of the researches of Charles Booth and his poverty map (Map 80), while cholera epidemics there were eradicated after John Snow demonstrated in 1855 that the disease was borne by water through mapping an outbreak and tracing the source of the contamination to a water pump (Map 79).

Technical changes in mapmaking both reflected contemporary developments and understandings, and also influenced later mapmakers and map-users. Thus the invention of printing by moveable type in the West and the production of printed maps for the first time (Map 13) helped to spread both an awareness of the value of cartography and also the influence of the messages that the maps conveyed. In the European Renaissance, globes became effective tools for showing the distribution of land around the world (Map 32). In the 18th century, Cassini's map of France (Map 57) was the first to be based on a triangulation of an entire country. Later, the Ordnance Survey in Britain (Map 61) set new international standards for national mapmaking, and the maps were used as the basis for much other mapping.

Do maps change the world? The answer is yes: maps change worlds of empires, kingdoms, local societies, and individuals, all over the world and at all times. They do this by selecting particular pieces of information to show—and items to omit or amend—and by conveying the material in ways appropriate for their purpose. *To the Ends of the Earth* is the story of just 100 of these maps.

Sarah Bendall
Emmanuel College, Cambridge

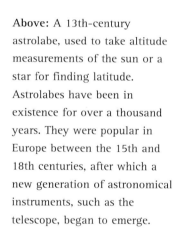

Above: A 13th-century astrolabe, used to take altitude measurements of the sun or a star for finding latitude. Astrolabes have been in existence for over a thousand years. They were popular in Europe between the 15th and 18th centuries, after which a new generation of astronomical instruments, such as the telescope, began to emerge.

THE ANCIENT WORLD: Representing the Landscape

Exactly how, when, why, and where the first maps came to be created is difficult to discover. Much of what was drawn in prehistoric and early historical times has not survived, so what we find today may not be wholly representative of what was once there. There are other problems for the modern observer. Maps made in prehistoric times—like those made today by non-literate or indigenous societies—cannot be accompanied by a title that explains the meaning of the drawing or that describes its content. However, we may be sure that in early times, just like today, maps were created for a variety of purposes and took a variety of forms. We may also be sure that, contrary to popular belief, of all the purposes to which maps have been put through the ages, the least important single purpose has been to find the way. Sea charts did not come into existence until the European Middle Ages, and topographical maps were not normally carried about by land travelers until the 18th century.

Human beings have always had the capacity to think spatially—this is *here*, that is *there*—even though not everybody chooses to express this understanding in mapmaking. Mental maps are almost certainly one of the oldest forms of human communication. Hunter-gatherers, for instance, catalogued the routes of the migratory animals they were following and the best places to hunt them down. Wandering tribesmen needed to know how they could cross deserts safely without dying of thirst. All these people would have carried a map of their territory in their head.

Modern archaeological research confirms that the notion of creating maps and plans developed independently in different areas of the world long before the first days of literacy. The earliest maps that survive bear witness to the cultures, priorities, and beliefs of their makers exactly as the maps we use today reflect our own preoccupations. As today, some maps were based on personal experience and familiarity with a local situation; others, such as cosmological maps, were works of imagination. Less familiar to modern readers will probably be the role of maps in prehistoric times, or in non-literate societies, in a religious context.

1 World Map, Babylon, c. 600 B.C.E.
Devised to show the world and its encircling earthly ocean according to the beliefs of its creators, this is one of the earliest surviving world maps. The map, which is incised on a clay tablet, is accompanied by an explanatory cuneiform text. Babylon itself is portrayed schematically. Eight triangles—four of which can still be seen here—stood for the islands the Babylonians believed were a bridge to the heavenly ocean beyond.

THE EARLIEST MAPS

One of the earliest surviving maps dates from around 6200 B.C.E. (see p.13). It comes from the prehistoric site of Çatal Hüyük, in Anatolia, where it was unearthed by archaeologist James Mellaart and his team in 1963. The map is the oldest known town plan in existence today. It is not a measured or surveyed plan, but a picture map, with some parts represented in plan (as if seen from a viewpoint vertically above the town) and some parts represented in profile (as if seen from ground level). The mixture of profile and plan is still characteristic of many modern maps. Look closely at a modern topographical map and you will notice some landscape features represented by map signs showing the profile of the feature, such as windmills or the crossed swords of a battlefield. Like the Çatal Hüyük map, not all modern maps are drawn to scale either—consider some modern tourist maps, for example. We call the Çatal Hüyük wall painting a map rather than a picture, or view, of the town because it includes an element of plan. When we define a drawing as a map or plan, providing it shows the correct distribution of features—that which is next to something in reality is shown next to it—the manner of drawing and the lack of a mathematical scale are unimportant.

2 Town Plan, Çatal Hüyük, Turkey, c. 6200 B.C.E. This wall painting, showing the streets and houses of a Neolithic settlement, is thought to be the oldest town plan in existence. Archaeologists believe that the plan was created for ritual purposes and had no practical function. It is one of the main reasons why the Turkish site is widely regarded as one of the most important archaeological finds of recent times.

Çatal Hüyük, Turkey
The settlement is Neolithic, consisting of a series of layers or levels, each featuring mud-brick dwellings and other buildings built next to each other and sharing common walls.

The Çatal Hüyük plan

The map found at Çatal Hüyük is a 9-foot-long (275 cm) wall painting (the original may have been even bigger, as only fragments have survived). It was discovered decorating the walls of a shrine on Level VII of the excavation. The plan shows some 80 buildings arranged on rising terraces, accurately capturing the beehive design of the settlement. In the distance, an erupting twin-coned volcano is drawn in profile rather than in plan. Its slopes are covered with incandescent volcanic bombs. Others are being thrown from the erupting cones, above which hovers a cloud of smoke and ashes. The fact that the volcano has two cones suggests that it is Hasan Dag, which stands at the eastern end of the Konya plain and is visible from Çatal Hüyük. This and other volcanic mountains of the area were important to the town's inhabitants since they were a source of obsidian, which was used in the making of tools, weapons, jewelry, mirrors, and other objects.

The Çatal Hüyük map is a sophisticated representation of this Neolithic settlement. What makes it even more remarkable is the fact that, as confirmed by radiocarbon dating, the map was created some 2,700 years before cuneiform writing was invented by the Sumerians in the Fertile Crescent between the Tigris and Euphrates Rivers in the fourth millennium B.C.E.

Rock art and petroglyphs

The inhabitants of Çatal Hüyük were not the only preliterate people to produce maps. One complex of rock carvings in Val Camonica, near Brescia in northern Italy, has been interpreted as showing fields and paths in plan. The carvings date from the middle Bronze Age (c. 1200 B.C.E.), with others superimposed in the Iron Age (c. 800 B.C.E.). The complex covers a rock surface measuring 13.5 x 7.5 feet (4.1 x 2.3 m). There is little to guide the modern observer's interpretation. Irregular lines may represent paths or streams. Various points and circles are likely to signify springs. Stippled rectangles may represent orchards, and empty rectangles other plots cultivated by the community. It is thought unlikely that the map represents any specific place, but rather "place" in general. It is believed that the carvings were made to invoke the gods' protection for the local community's farmsteads and crops. Such carvings suggest that the mapping impulse played an important role in early human societies, and that the maps these societies produced were linked to their religious beliefs.

Petroglyphs have been discovered wherever rock art has been identified by archaeologists, in places as far removed as Idaho and southern Australia. The Australian Yunta engravings, created between c. 13000 and 11000 B.C.E., are also thought to have had a spiritual rather than geographical significance. As well as representing elements of the landscape, the markings are thought to relate to the Dreamtime, the Aboriginal myth of creation. The engravings show abstract circular signs as well as the tracks of birds and animals.

The Idaho site, christened Map Rock by early white settlers, is one of the oldest rock maps in North America, dating from c. 10000 B.C.E. (see p.14). It is situated on the northern side of the Snake River, and some experts believe that its carvings are a map of the river basin. They argue that the line running down the center of the rock represents the course of the Snake River, the lines off it signifying its tributaries, and the round figures show mountains. However, interpretations of rock art are highly subjective, as the individual markings could have been made on different occasions, perhaps widely separated in time.

MAPS IN MESOPOTAMIA

Around 6,000 years ago, the first centralized, hierarchical communities began to develop into larger states and eventually, in some cases, into the world's original empires. It was in early

3 Map Rock, Idaho, USA, c. 10000 B.C.E.
Early Native Americans may have made these markings, on a basalt rock on the north bank of the Snake River, over the course of hundreds of years. It is thought by some that the carvings could be a map showing the course of the river through its basin. The markings are likely to have had a votive role.

urban-based societies such as these—in Egypt, China, the Indus Valley, the central Andes, Mesoamerica, and southwest Nigeria—that more detailed large-scale plans developed. In Mesopotamia, between 4000 and 3000 B.C.E., an urban-based society began to emerge, paralleled by the development of an organized system of agriculture. Field patterns, irrigation systems, and, most importantly, landholdings were key elements of the economy and hence the subject of most of the early maps the Mesopotamians are known to have produced. Such maps are early cadastral plans (maps showing property boundaries), serving, perhaps, as a kind of title deed where a visual record was deemed more useful and clear-cut than any written account of the boundaries of a property. In a region as dry as Mesopotamia, the position and maintenance of irrigation canals was also of particular interest, for cultivation without irrigation was hazardous and crop choice was limited.

Clay tablets

The medium these Mesopotamian cartographers employed was the inscribed clay tablet, of which thousands of examples have been found dating from around the time when Sargon I was founding the kingdom of Akkad in about 2350 B.C.E. to the time of Babylonian supremacy (c. 612–539 B.C.E.). One of the earliest examples of a map of this kind was discovered in 1930 by archaeologists digging at Yorghan Tepe, near Kirkuk, some 200 miles (320 km) north of Babylon. The likely date of the tablet is around 2300 B.C.E. Small enough to fit into the palm of the hand, its surface is inscribed with a map of the area around Gazur (later Nuzi) bounded by two ranges of hills—perhaps the Zagros Mountains on the border between Iran and Iraq—with a waterway, either the River Euphrates or perhaps an irrigation canal, running through the middle. Cuneiform inscriptions highlight certain features and places: in the center, for instance, the size of a particular plot of land is stated, while its owner is identified by name as Azala. Circles indicate north, east, and west at the left, top, and bottom of the tablet respectively. This is the earliest known example of map orientation to have survived. The signs employed to represent cultural and natural features—circles for the three towns on the map, double rows of images resembling bells for mountains, and double lines for waterways—are notable for their clarity.

The purpose of a later map tablet, dating from around 1500 B.C.E., also appears to have been the delineation of landholdings. It is a plan of Nippur, the Sumerian capital and religious center just south of Babylon, recording the landholdings of the religious and political elites of the day. It shows what seem to be fields around the bend in the River Euphrates, with irrigation canals marking the boundaries between the various estates.

Another map tablet of around the same date, showing a plan of approximately a quarter of Nippur, is even more detailed. It shows the Euphrates, two canals, and the city's major buildings, including the Temple of Enlil in its own square, two large storehouses by the river, and the city walls with seven named gates. A noteworthy feature of the map is that measurements in cubits are given for some of the buildings shown, which implies that the plan as a whole may well have been drawn to scale, based on a measured survey, and probably commissioned to help in building new fortifications. If this supposition is correct, it is an example of the introduction by the Mesopotamians of maps drawn to scale. The first known example of a large-scale Mesopotamian plan with an indication of scale is slightly earlier than this. The plan, which dates from around 2100 B.C.E., is incised on a statue of King Gudea, ruler of Lagash in Sumeria, on the robe lying across the king's lap. Traces of a scale bar can be seen clearly on a broken corner.

A number of Mesopotamian clay tablets, dating from between 2500 and 2200 B.C.E., are inscribed with cuneiform lists of place names, rivers, and mountains. Although there is no firm evidence to connect the existence of these tablets directly with mapmaking, the likelihood is that there was some link between the two. Such tablets may have been used in teaching, as well as perhaps for military purposes. Under Sargon, the Akkadians were launching military expeditions westward from around 2330 B.C.E., and this would account for the inclusion of places as far west as the Mediterranean in the lists.

The Babylonian world map

Mesopotamian mapping reached the height of its achievement with the rise of Babylon to political, military, and eventually imperial prominence. One of the most celebrated cities of antiquity, Babylon reached the apogee of its power as the capital of the mighty empire carved out by King Nebuchadnezzar around 600 B.C.E. At that time, it was the world's largest city, covering 2,500 acres (10 sq km) and containing imposing temples, ziggurats, and palaces, as well as the famous Hanging Gardens, later acclaimed as one of the seven wonders of the ancient world. One clay tablet, dating from around 500 B.C.E., maps the city, locating the Temple of Marduk and the route of a processional path that led through the Ishtar Gate and on to a smaller temple located outside the city's walls.

The Babylonians were the first to formulate many of our most familiar mathematical ideas, such as the notion that a circle is divided into 360 degrees. We are reminded of this when we look at a remarkable clay tablet map of the Babylonian world, which has been boldly inscribed with a perfect circle (see p.10). The tablet dates from around 600 B.C.E., but the map was probably a copy of something made some two or three centuries earlier that has not survived.

The map is small, as it occupies only two-thirds of the surface of a clay tablet that itself measures no more than 5 x 3 inches (125 x 75 mm). The rest of the tablet is taken up by an explanatory text, which confirms that the map was intended by its makers to show the whole of the world. This is depicted as a flat disc surrounded by, or floating on, an ocean. Beyond the ocean lie outer regions or islands, marked with triangles, originally probably eight in number, attached to which are figures thought to refer to the distances between them. Each of the islands is described in some detail in the cuneiform inscriptions accompanying the map. The "islands of transition," as they were termed, were beyond human reach. One of them was a place of light "brighter than that of sunset or stars," while another, to the north, was shrouded in perpetual darkness. A "horned bull" on the sixth island attacked all visitors.

North is at the top of the map. A rectangle near the center stands for the city of Babylon. Around it are eight circles representing other cities. Assyria is to the right with what may be Uratu (eastern Turkey and Armenia) above it. Vertical parallel lines, possibly standing for the Euphrates, extend from the mountains at the top to the marshlands of lower Mesopotamia— these are depicted by horizontal parallel lines at the bottom of the circle—and eventually the sea at the bottom. The whole representation offers an insight into the Babylonians' conception of their universe. Although there is some evidence from other sources of Mesopotamian traders having reached westward into the Mediterranean and eastward into the Indian Ocean, nothing they saw there is featured on this particular map, which concerns an explicitly Babylon-centric cosmos.

Cosmological maps

The Babylonian world map thus reflects the spiritual beliefs of the people who created it, most notably in the way the world is depicted as circular and surrounded by water. Fitting in with what we know of Babylonian cosmology, the eight islands are the links between Earth's seas, the heavenly ocean, and the animal constellations that inhabit it. Yet the Babylonians were by no means the first or the only society to link the notions of cosmology and mapping. Cosmological maps were devised by ancient cultures all over the world—from Egypt and India to North and South America—and they also feature significantly in medieval Christianity and in Islam.

Cosmological maps were intended to be statements of belief. They are symbolic of the worldview of the cultures of their creators, and are reflections of the religious beliefs of the

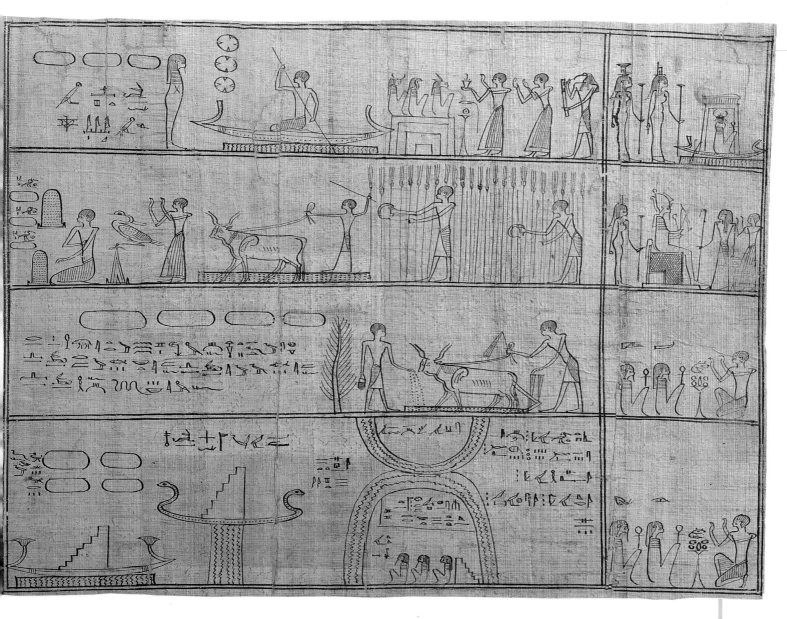

societies of their times. In North Africa, for instance, a labyrinth-like rock painting charts the path through life to death and on into the afterlife, while the prehistoric Triora Stela, from Brescia in Italy, depicts heaven at the top, the earthly world in the center, and the world of the dead at the bottom.

In ancient Egypt, cosmographical maps were common, particularly in royal tombs of the 18th to the 20th dynasties (c. 1550–1070 B.C.E.), where coffins were frequently decorated with pictorial plans on the inside of their lid, to assist the newly dead in their journey to the afterlife.

MAPS AND THE EGYPTIANS

The level of sophistication reached by the civilization of ancient Egypt can be argued to have rivaled that of ancient Mesopotamia, but maps and cartography do not seem to have played the same role in the two societies. The Babylonians often used their mapmaking skills for practical, organizational purposes, whereas the Egyptians seem to have focused much of their attention on creating cosmological depictions of the Earth and the heavens, portrayals of mythical lands, and, as decoration on coffins and tombs, elaborate routes to the afterlife through the land of the dead. The painted coffins of El-Bersha, for example, which date from around 2000 B.C.E., illustrate

4 Journey to the Afterlife, Egypt, c. 1400 B.C.E.
This illustration from a Book of the Dead is typical of the religious geography of the ancient Egyptians. The papyrus shows a dead soul on its way to reach the kingdom of the god Osiris, where a plot of fertile land stands ready to be tilled. Other Egyptian tomb artifacts illustrate two different ways of reaching the gateway to the underworld, one by water and the other by land.

different ways of reaching the gateway to the underworld.

The ancient Egyptians, just like the Mesopotamians, regarded accurate land measurement as important, and they had good reason to do so. Much of the land inundated by the annual Nile floods had to be resurveyed once the river had receded in order to re-establish the boundaries between farms, estates, and other properties. Professional surveyors were employed for the task, for it was vital for Egypt's ruling pharaohs to know who owned what land and how much of it, not least of all for tax purposes. No survey maps seem to have survived from the dynastic period: the oldest ones still in existence date from the time of the Ptolemys (304–30 B.C.E.).

However, Apollonius of Rhodes, who became librarian at Alexandria in 196 B.C.E., reported in his *Argonautica* on the mapmaking skills of the inhabitants of Colchis, an Egyptian colony since the time of Rameses II (c. 1290–1224 B.C.E.):

" *They preserve the writings of their fathers, graven on pillars, whereon are marked all the ways and the limits of sea and land as ye journey on all sides round.* "

Most of the topographical maps that have survived are drawn in the usual combination of plan and profile. The most common subjects are gardens, in which sycamores and date palms surround paths, with orchards or vineyards indicated outside the garden boundaries and people and animals represented in elevation.

Planning a pharaoh's tomb

Although maps of some buildings dating from between 1500 and 1000 B.C.E. combine elements of elevation and plan, most of the great pyramids and temples seem to have been

5 **Turin Papyrus, Egypt, c. 1300 B.C.E.** This painted papyrus, which has survived only in fragments, is a map of the hills between the River Nile and the Red Sea, and shows the location of the gold and silver mines in the area. It has been suggested that the colored areas indicate different geological outcrops, which would make this the world's first known geological map.

erected without using such drawings. There are, however, a few noteworthy exceptions. One of these was the tomb of Rameses IV, which had not been completed by the time of the pharaoh's death in 1160 B.C.E. Two plans survive, one drawn on papyrus and the other on a broken pottery shard. The use of papyrus, a writing material made from a reedlike plant cultivated in the Nile Delta, was unique to the ancient Egyptians and more practical for plan-making than the clay tablets favored by their contemporaries elsewhere in the Middle and Near East.

The version of the plan on papyrus, though incomplete and somewhat damaged at its bottom, seems to show the tomb's internal arrangement after the sudden death of Rameses had forced a hasty rethink of the layout of the interior. The major task seems to have involved adapting the large hall that would normally have preceded the burial chamber into a setting worthy of housing the actual sarcophagus. The plan sets the tomb in an overall geographical context: the surrounding desert is outlined in red and hatched with dotted lines on a black background. Though the sequence of rooms within the tomb corresponds to that of the actual building, there was no attempt to draw them to scale or to show their shape. The plan, it seems, was not intended as a record of the finished work, but as a working guide, probably devised to help the workmen engaged in refashioning the tomb. Evidence to support this interpretation comes from the plan's annotations, which give information about room sizes and measurements as well as recording how quickly the reconstruction work was progressing.

The Turin papyrus

Like unbaked clay tablets, papyrus is a fragile medium, which may be why so few ancient Egyptian plans and maps have survived. The Turin papyrus is an exception. It is a topographic map of a region in Nubia, and was painted in around 1300 B.C.E. It portrays, among other things, a settlement associated with gold- and silver-mining, the location of the mines themselves, and the roads linking the River Nile with the coast of the Red Sea. The map was discovered in fragments around 1824, when it passed into the possession of Bernardino Drovetti, an Italian diplomat and antiquarian who was serving as French consul in Egypt. With the many other antiquities he collected, the papyrus formed part of the collection that Drovetti donated to found the Egyptian Museum in Turin, hence its name. It is still kept there today.

The map was created by Amennakhte, Scribe of the Tomb to Rameses IV. The pharaoh was planning a quarrying expedition to Wadi Hammamat, in the eastern desert, with the purpose of obtaining blocks of sandstone for a giant statue of himself, and the map was intended as a guide to the area. Its most prominent features are the winding wadi (dry watercourse) and its confluence with the neighboring wadis of Atalla and el-Sid, the surrounding hills—and the location of gold deposits within them—the stone quarry, and the gold mine and settlement of Bir Umm Fawakkhir. Annotations identify the features shown on the map, and describe to where the various tracks along the wadi led, the distance between the quarry and the mines, and the sizes of the stone blocks that were quarried. The map is orientated to the south, where the source of the Nile lies.

It has been suggested by some scholars that the papyrus is actually a geological map. If this supposition is correct, it would be the earliest that is known. It has been argued that the map accurately distinguishes two main rock types by coloring the hills either black (for schists) or pink (for granites), and the brown, green, and white dots within the main wadi may also have some geological significance. Other scholars, however, dispute this view and say that there is no hard evidence to show that the map was compiled in this way. They also argue that the use of colors, although original, was more likely to have been a means of helping to locate landmarks or for decoration than for showing anything of geological import.

Tomb of Rameses IV
c. 1160 B.C.E.,
Valley of the Kings,
Luxor, Egypt
The chamber of the
sarcophagus (shown here)
was refashioned after
Rameses' sudden death with
the help of a plan drawn on
papyrus as a working guide.

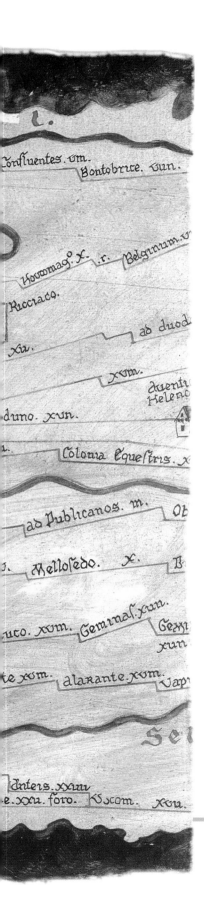

THE CLASSICAL WORLD: Creating a Science

C artography owes an immeasurable debt to the ancient Greeks. Most of the Ionian philosophers and mathematicians of the great age of Classical Greece, between c. 600 B.C.E. and 200 B.C.E., largely concerned themselves with theoretical rather than practical mapmaking, and laid the foundations on which subsequent generations of Western cartographers were to build. The contribution of the Greeks was considerable: for example, they first put forward the idea of the world being a sphere.

The shape of the Earth had been a matter of speculation for more than a thousand years. The Babylonians believed in a flat, circular Earth floating on, and surrounded by, ocean, while the ancient Egyptians believed the world was egg-shaped. Anaximander of Thales (c. 611–546 B.C.E.), later hailed by the Greeks as their first mapmaker, probably believed the world to be circular, but no details of the world map it is claimed that he drew have survived. The mathematician Pythagoras (588–c. 500 B.C.E.) envisaged the Earth as a perfect sphere, which Parmenides of Elea (born c. 515 B.C.E.) divided into five zones: one hot, two temperate, and two cold.

Eratosthenes (276–194 B.C.E.), a Greek who was appointed tutor to the son of the Egyptian king Ptolemy III Euergetes and who was subsequently made director of the library at Alexandria, made a great contribution to the cartography of the day. He was probably the first man to measure the circumference of the Earth. The result he achieved was astonishingly accurate, distorted only by his acceptance of Pythagoras' belief that the world was a perfect sphere, whereas it is really a slightly flattened one.

One problem in making a full assessment of the Greek contribution to cartography is the scarcity of original source material. Few, if any, maps made during the age of Classical Greece have survived. This is largely because the maps that were undoubtedly created were drawn on perishable or reusable materials. Maps were normally painted on wood, paper, papyrus, or cloth, or less often engraved on bronze. The maps that do exist are mostly medieval copies.

Not only the maps themselves have been lost. Eratosthenes, for instance, wrote two lost texts—the *Measurement of the Earth* and *Geographica*—in which he outlined the methods he developed to measure the circumference of the Earth and described how to make a map of the world. Fortunately, the writings of men such as Herodotus (c. 495–425 B.C.E.), a historian with a deep interest in geography, and Strabo (c. 64 B.C.E.–

6 Peutinger Table, Rome, 4th century C.E. Probably copied by a Franco-German monk in around C.E. 1275 from a 4th-century original, the so-called Peutinger Table is a detailed route map of the entire Roman Empire. This section is devoted to Gaul (modern France). Although only this copy survives, the map is invaluable in assessing Roman cartographic accomplishment.

C.E. 21) have survived, and help us to build up a picture of the early Greek mapmakers and their achievements. Herodotus was probably the first author in Western literature to refer specifically to a map, when, in his *Histories*, he wrote of "a bronze tablet with an engraving of the whole world with all its rivers and seas." For his part, Strabo wrote a 17-volume *Geographica*, which not only included an informative assessment of earlier geographical writers and mapmakers, but also provided a vivid verbal picture of the world as it was known at that time.

It is only through Strabo's writings and those of Cleomedes, another Greek scholar of the day, that our knowledge of Eratosthenes' mapping has survived. We know that he presented a world map to the Egyptian court, and that the map was an attempt to draw the world to scale. The map used an orthogonal projection, in which the known world was flattened to fit latitudes and longitudes at right angles to each other. The map's influence on Greek and Roman cartographers was profound, even among later scholars who felt that there was a more scientific way of dividing up the world than the arbitrary network of parallels and meridians that Eratosthenes had used.

PTOLEMY AND HIS LEGACY

Claudius Ptolemaeus, better known today as Ptolemy, lived from around C.E. 90 to about C.E. 168. His first name is Roman, but his second is that of the Macedonian dynasty that ruled Egypt from the time of Alexander the Great until its conquest by Rome. Little is known about Ptolemy's life. It is thought that he was born and raised in Upper Egypt but that he spent most of his adult career in Alexandria, which, with its renowned library, had become the intellectual capital of the Hellenic world.

Having established the mathematical groundwork for his description of the Earth in his *Almagest*, Ptolemy started to compile what became his greatest work, the eight-volume *Geographike Hyphegesis*, better known today as *Geography*. The work is both a treatise on the mathematical construction of maps (a solution to the problem of depicting a sphere on a flat surface) and a gazetteer of some 8,000 place names and geographical features, together with their coordinates of latitude and longitude. The coordinates enabled each place to be plotted on a map in relation to their actual position in the world as measured by astronomical observation. The treatise was given in book one, while the coordinate tables filled books two to seven. The work therefore gave instructions for the compilation of mathematically based maps.

The prime task of the cartographer, Ptolemy wrote, was to survey the world to scale:

> " *The task of geography is to survey the whole in its just proportions, as one would the entire head.* "

Ptolemy employed two map projections: a conic one, in which the parallels of latitude are shown as concentric circular arcs; and a pseudoconic projection, in which the lines of longitude are circular rather than straight. His maps were orientated with north at the top. Book eight of the *Geography* is believed to have contained maps of individual regions, with ten maps of Europe, four of Africa, and twelve of Asia. None of the maps has survived, and scholars disagree about whether Ptolemy actually drew the maps to which he refers or just gave instructions about how to compile them. What did survive was a copy of the work—without maps—which was translated into Arabic in the 9th century. The work then found its way to Byzantium, where maps were drawn, but was not made available to the Latin West until 1406. Since then, Ptolemy's ideas for map construction have dominated cartography, particularly topographical and world mapping, right up to the present (see p.24).

Ptolemy's *Geography*, Latin manuscript, c. 1406
The text of Ptolemy's great work was translated into Latin in 1406. This folio lists the coordinates, latitude and longitude, for locations in Crete, alongside a map of the island.

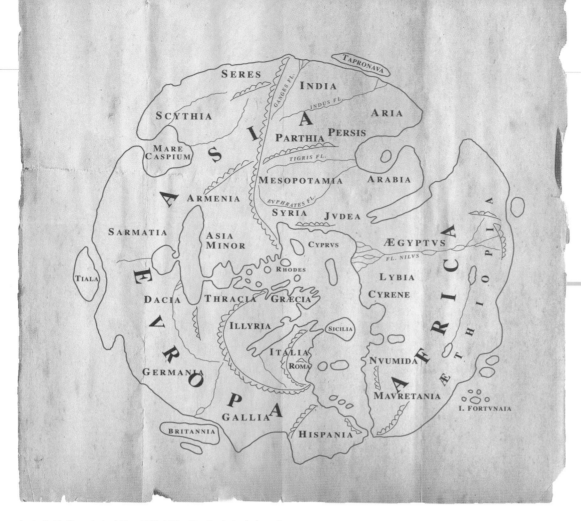

Map labels: TAPRONAVA · SERES · INDIA · SCYTHIA · GANGES FL. · INDUS FL. · ARIA · PARTHIA · PERSIS · MARE CASPIUM · ASIA · TIGRIS FL. · MESOPOTAMIA · ARABIA · ARMENIA · EVPHRATES FL. · SYRIA · JVDEA · SARMATIA · ASIA MINOR · CYPRVS · ÆGYPTVS · FL. NILVS · ETHIOPIA · TIALA · RHODES · LYBIA · CYRENE · DACIA · THRACIA · GRÆCIA · AFRICA · ILLYRIA · SICILIA · ITALIA · ROMA · NVMIDA · GERMANIA · EVROPA · MAVRETANIA · I. FORTVNAIA · GALLIA · BRITANNIA · HISPANIA

7 Reconstruction of Agrippa's World Map, Rome, c. 7 B.C.E.

The earliest known Roman world map, displaying the full might of the empire, was largely the work of Marcus Vipsanius Agrippa, son-in-law of Emperor Augustus. It was the most ambitious attempt yet made to represent so vast a territory with some degree of geographical precision. Although copies of Agrippa's map were taken to all of the great cities of the Roman Empire, not a single copy is known to have survived. However, detailed descriptions of what the map showed appeared in the elder Pliny's *Natural History*, in which the most common references are to the size of provinces or groups of provinces, though land and sea measurements are also cited. This reconstruction is orientated with the east at the top.

MAPS AND THE ROMANS

While Greek mapmakers were providing the theoretical basis for subsequent Western cartography, their Roman counterparts were considerably more pragmatic in their mapmaking activities. They were less concerned with elaborate mathematical calculations than with conveying information in forms that best suited their political, military, or administrative purposes.

Julius Caesar (100–44 B.C.E.) commissioned what was perhaps the earliest Roman world map, but he did not live to see it completed. In 44 B.C.E., the year of his assassination, he asked four Greek cartographers living in Rome to survey the four quarters of the known world. After Caesar's death, the task of seeing the survey through to completion—it was to take more than 30 years—fell to the military commander Marcus Vipsanius Agrippa (64–12 B.C.E.), one of the earliest supporters of the young Augustus (63 B.C.E.–C.E. 14), then known as Octavian, in his struggle to establish himself as Caesar's heir. Although Agrippa also did not live to see the survey completed—Augustus took over the supervision of the final stages—the map bore Agrippa's name as a fitting memorial to the service he had given Augustus and the empire.

Augustus' reasons for wanting a map of the world were as much the product of political calculation as they were of his interest in geography and the new territories into which his empire was expanding. He was committed to founding new colonies in southern Europe and North Africa to provide land for veterans of the civil wars. Such a map, which he intended for public display in Rome, would supply an instant reference to the location of these colonies. Augustus was also determined to build up the notion of Rome's imperial benevolence among the myriad peoples he now ruled. He might well have reasoned that Agrippa's map could be a powerful propaganda tool to enhance this image. When the map was finally displayed (it was either carved in, or painted on, marble and fixed to the wall of a portico along the Via Lata) the extent of the empire could be appreciated at a glance. The map was probably rectangular in shape,

8 Ptolemy's World Map, Egypt, c. C.E. 150

Although we do not know what this and other maps Ptolemy himself might have drawn actually looked like, early Renaissance cartographers were able to work from his writings—rediscovered by the Arabs and later translated into Latin by the Byzantines—to reconstruct his work, such as this map of the world. The Mediterranean region, parts of Western Europe, and the Near East look much the same as they do on modern maps, but there were omissions and errors, giving us an insight into the Classical Greek worldview. Ptolemy knew nothing of the existence of the Americas, while he showed the Indian Ocean as a vast enclosed sea with Taprobana (Sri Lanka) large enough visually to overwhelm the entire Indian subcontinent. Renaissance cartographers did not realize that Ptolemy had underestimated the circumference of the Earth by about a quarter. Ptolemy's mathematical ideas, as displayed in this map, dominated much of geographic and cartographic thought from around 1400 to the present day.

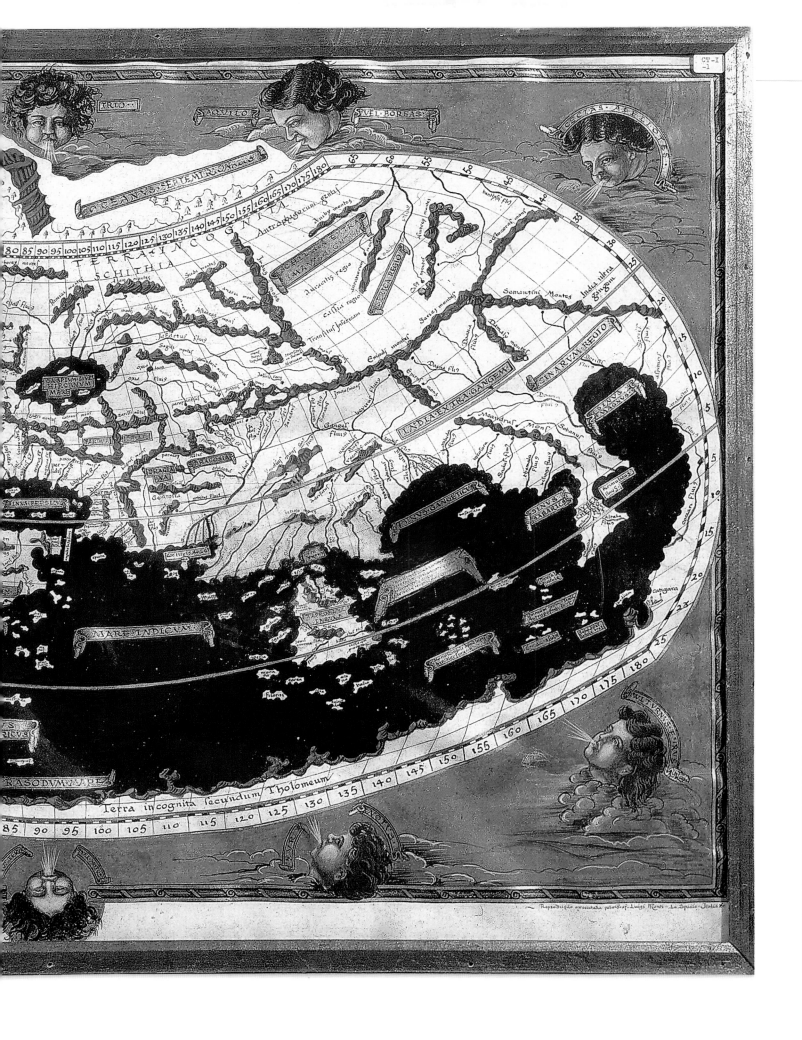

9 Forma Urbis Romae, Rome, C.E. 203–208 Carved in marble, this monumental map of imperial Rome is the only major Roman city plan known to survive, albeit in fragmentary form. This particular fragment shows part of the Subura neighborhood, a particularly insalubrious part of the city. In addition to the main street, known as Clivus Suburnus, and the Porticus of Livia, shops, apartment blocks, baths, and even a possible brothel are marked. The plan's scale and detail provide a unique picture of Rome.

with north at the top. Neither the map—nor any of the copies that were supposedly made of it to be displayed in other imperial cities—have survived. However, there are detailed descriptions of what it showed in the elder Pliny's celebrated *Natural History*, although it is uncertain whether Pliny saw the map or relied on the commentary Agrippa had drawn up to accompany it.

Land surveying

Although there are few records of maps from early Roman times, it is clear that mapmaking was from the start dominated by the demand for accurate, detailed land surveys. The usual practice was for two copies of all such maps to be made: one for the state archives in Rome and the other for the local community. As Roman rule expanded to encompass most of the known world, these surveys became ever more important. Conquered land was divided into holdings for veterans as a means of controlling the regions. The division of land into rectangles—in towns, *insulae* (blocks) of varying size, and in the country, square *centuriae* (centuries) mostly measuring 2,400 by 2,400 Roman feet—was an essential preliminary to establishing an effective system of administration.

Textbooks about how to conduct such surveys started to appear in the 1st century B.C.E.; the earliest Roman survey maps known to survive date from between 167 B.C.E. and 164 B.C.E. In about C.E. 350, a collection of the most important surveys was brought together in the *Corpus Agrimensarum* (Body [of writings] of the Land Measurers). The original manuscripts were written on papyrus and have long since vanished, but a copy, made around C.E. 800, is thought to be substantially accurate. Surveying continued to be central to Roman mapmaking right up to the collapse of the western part of the empire in the late 5th century C.E.

City plans

Roman cartographers were also concerned with mapping their cities. An outstanding example of such a map is the Forma Urbis Romae, an incredibly detailed survey of Rome, which was carved on 150 marble slabs some time between C.E. 203 and C.E. 208, when Septimius Severus was emperor. Whether or not this was the first large-scale map of Rome to be created is uncertain, for there is evidence to suggest that Emperor Vespasian and his son Titus had ordered the measuring of the city in C.E. 74. However, the Forma Urbis Romae was conceived on a truly heroic scale, as we can tell from the 1,200 or so fragments that have survived.

When completed, the map measured 59 x 43 feet (18 x 13 m). It hung on an outer wall of a library attached to Vespasian's Temple of Peace, which Septimius Severus had ordered to be restored after a disastrous fire in C.E. 191. The purpose of the map was simple: to inspire onlookers with awe at the splendor and sheer size of the imperial capital. It was sufficiently detailed to enable the accurate recording of every public and private building throughout the city. The orientation of the map is approximately southeast and it is centered on the Capitoline Hill. The scale averages around 1:300, although some public buildings are disproportionately large to emphasize their importance. Certain temples are colored in red for the same reason.

A route map

The old saying "all roads lead to Rome" is, of course, not strictly true, but it is the case that the Romans were the pre-eminent road-builders of the Classical world. Despite the relative complexity of the Roman road system, travelers would have relied on word of mouth and local guides for finding their way. Some might have had access to an itinerary: a list of places from a starting point in correct sequence.

The Peutinger Table is a 13th-century copy of a Roman route map which probably dated from the 4th century C.E. (see p.20). The map shows routes mapped out from a collection of travelers' itineraries. Most of the town names on the map are in the ablative or accusative ("from Rome" or "to Rome"), showing that they were copied from written descriptions of the route from one place to another. Perhaps the map was created purely for display, as a vanity piece, by someone who collected together all the itineraries they could find, irrespective of age.

The map takes its name from Konrad Peutinger (1465–1547), an Augsburg antiquarian who inherited it from a librarian named Konrad Bickel in 1508. Bickel was vague about the map's provenance, but Peutinger recognized it for what it was and appreciated its significance as a unique record of the Roman world.

What makes the map immediately striking is its odd size and shape. It consists of eleven sheets of parchment—there was a twelfth probably showing Spain, Portugal, and western Britain, but this has long been lost—whose total width is about twenty times their height. Nor is there any consistent shape or scale. As a result, distances from north to south are extremely short, while distances from east to west invariably appear too long. It is possible that the map's odd dimensions are due to the fact that the original 4th-century map was a copy of an earlier papyrus roll. Although such rolls could be of considerable length, their width was limited.

The map covers the Roman Empire, the Near East, India as far as the River Ganges, and Sri Lanka. Even China is mentioned. No fewer than 555 cities and 3,500 other place names are marked. Routes are shown as continuous lines with numerals indicating distances between places on them. The system of markings the mapmaker employed is extremely elaborate. Signs represent everything from towns to important villas, temples, spas, granaries, and lighthouses.

MAPS IN BYZANTIUM

10 Mosaic Map, Madaba, Jordan, c. C.E. 550 Though it survives only in part, the mosaic map laid out on the floor of a village church near Amman in Jordan is the oldest map of the Bible lands known to survive. Its designers were Byzantines. This illustration is of Jerusalem. The city is seen as though being viewed from the air and in considerable detail. Religious and public buildings have red gabled roofs.

A systematic study has not been made of the part played by the Byzantines in cartographic history. However, Byzantium certainly played a significant part in preserving existing knowledge: it was from Byzantium, for instance, that knowledge of Ptolemy's work made its way to the late medieval West.

It is also from the Byzantine world that one of the odder interpretations of the structure of the Earth has come down to us. Cosmos Indiopleustes (fl. C.E. 540) was a merchant from Alexandria who traveled widely—possibly as far as India, which is what his last name implies—before settling down to write his twelve-volume *Christian Topography* and two other books that have been lost. Cosmos' ideas about a rectangular flat Earth covered by a sort of dome or box were entirely his own, and found no favor with his contemporaries. Copies of his *Christian Topography* were made in Byzantine times, but more as a curiosity than from conviction. Western European interest in his work dates only from the 17th century, when it was held up to ridicule as an example of misguided medieval beliefs.

The Madaba mosaic

The largest and most detailed map from Byzantine times known to survive is the Madaba mosaic, which was created in the Greek Orthodox Church of St. George in Madaba, Jordan, between C.E. 542 and the death of the Emperor Justinian 23 years later. It was rediscovered in 1884 and, although only parts of it are intact, there is enough to amaze any viewer.

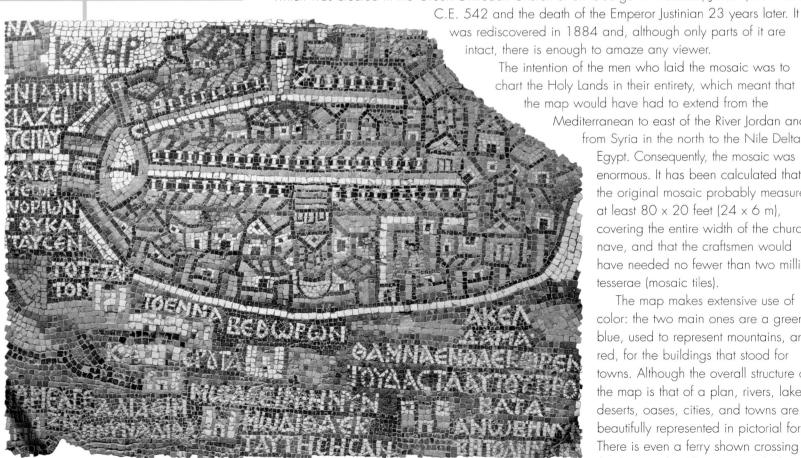

The intention of the men who laid the mosaic was to chart the Holy Lands in their entirety, which meant that the map would have had to extend from the Mediterranean to east of the River Jordan and from Syria in the north to the Nile Delta in Egypt. Consequently, the mosaic was enormous. It has been calculated that the original mosaic probably measured at least 80 x 20 feet (24 x 6 m), covering the entire width of the church's nave, and that the craftsmen would have needed no fewer than two million tesserae (mosaic tiles).

The map makes extensive use of color: the two main ones are a greenish blue, used to represent mountains, and red, for the buildings that stood for towns. Although the overall structure of the map is that of a plan, rivers, lakes, deserts, oases, cities, and towns are all beautifully represented in pictorial form. There is even a ferry shown crossing the River Jordan. The way in which Jerusalem is treated is particularly interesting. Its size was deliberately exaggerated, no doubt to make the point that, for Christians of the time and long afterward, it was one of the centers of the spiritual world. It is depicted as an oval walled city, and identifiable features include the Church

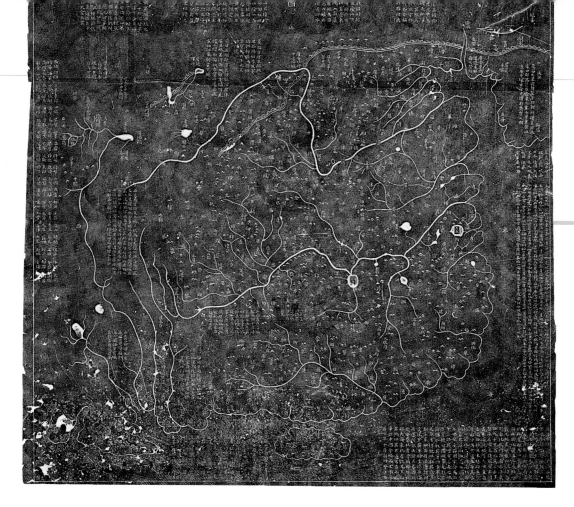

Stone Map, China, C.E. 1136
This map of the Chinese Empire, known as the Yu Ji Tu (Map of the Tracks of Yu), is highly detailed. Note in particular the grid squares overlaying the map, which indicate the scale. The writing lists place names of cities, towns, rivers, lakes, and mountains. Maps such as this were highly practical tools in the administration of the giant empire.

of the Holy Sepulcher, the New Church of the Theotokos (Greek for "Virgin Mary"), the Damascus Gate, and the public baths.

CARTOGRAPHY IN CHINA

The first Chinese map known to us dates from around 2100 B.C.E., while the earliest known survey of the country is thought to date from the 6th century B.C.E. During the period of Han domination, from 206 B.C.E. to C.E. 220, the Chinese established a tradition of mathematical mapping centered on an appreciation of the importance of scale.

Some early Chinese maps were created for military use, while others—land surveys, topographical maps—were drawn as a means of helping the ruling dynasty and its civil servants to administer the empire. One of the three maps discovered during the excavations of a Han period tomb at Mawangdu in the suburbs of Changsha City, Hunan, between 1972 and 1974, is believed to have fulfilled a military function. It was prepared for the ruler of Ch'ang-sha (modern Hunan) in around 168 B.C.E. It and its two companions are thought to be the earliest surviving maps in the world to have been compiled on the basis of field surveys.

From the 3rd century C.E., it became standard practice to itemize information such as distance and direction, together with descriptions of physical features, in text either on the maps or in accompanying pamphlets. These highly practical "maps with explanations," as they became known, were issued to Chinese imperial officials to help them familiarize themselves with the areas they were being sent to govern, making the maps key administrative tools.

It was Pei Xian (C.E. 224–271), one of the most noted scholars of the day who became Minister of Works in C.E. 267, who was responsible for establishing many of the principles that governed Chinese mapmaking. In his *Six Laws of Mapmaking*, Pei Xian stressed the importance of scale, location, distance, elevation, and gradient, as well as the need for careful consideration of accurate measurement and close attention to geographical detail.

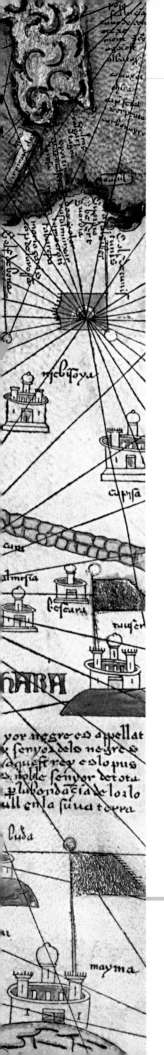

THE MEDIEVAL WORLD:
There Be Monsters?

Medieval society as a whole was not map-conscious. With the dawning of the Middle Ages, a new cultural world was coming into being: one in which, for very good reasons, the concept of mapping as it had existed in Greek and Roman times, gradually vanished. It is clear, for instance, that, although some Roman world maps survived into the early medieval period, the topographic skills behind them were slowly being lost as what survived of classical tradition and knowledge gradually became more and more corrupted. Such knowledge and skills were not to be rediscovered until the 15th and 16th centuries.

The last large-scale plan to be created according to the precepts of Roman surveying dates from 820 and comes from St. Gall in Switzerland. The plan, which was of an idealized model abbey, was drawn to scale and showed the entire monastic complex, with single lines for walls and each room or building carefully labeled according to its purpose. It was just this sort of skill that was gradually lost over the medieval period.

In medieval Europe, maps were rarities; it has been argued that they were almost unknown. They were certainly not a part of everyday life. From maps of the world to maps of a single locality, such maps that were created were produced in response to specific circumstances and for particular purposes. The navigational charts prepared from around 1270 onward for use by seafarers—firstly in the Mediterranean and then further north—are classic examples of such a response. So, too, are the 36 local maps that we know were produced in the British Isles during the medieval period. In each instance, the driving factor behind their creation was a local need.

12 The Catalan Atlas, Abraham Cresques, Majorca, 1375

In his great atlas, the celebrated Jewish mapmaker Abraham Cresques drew on the navigational charts and world maps of his times to produce a worldview far in advance of anything preceding it. The African king holding a scepter is Musse Melly (Mansa Musa), "the richest and most distinguished ruler of his whole region on account of the great quantity of gold that is found in his land."

The medieval worldview

Where do the great medieval world maps—the celebrated *mappae mundi* (the name comes from the Latin *mappa*, meaning "cloth," and *mundus*, meaning "the world")—created in northwestern Europe fit into this argument? The first thing to be understood is that it is an error to think of them as primarily geographic at all. The draughtsmen who created the Anglo-Saxon world map (see p.36), for instance, had no idea of the actual shape of any part of the world. The driving force behind their creation was religious and philosophical.

This, after all, was the great Age of Faith, and the mapmakers—most of them were churchmen or had been instructed by Church

Illumination depicting a
monk writing, *Vitae
Sanctorum*, Portugal,
12th century
In the Middle Ages the
monasteries were the central
storehouses of knowledge,
attracting the greatest scholars
of the day. Most mapmakers
were churchmen, and the
maps they created reflected
their belief in divine
order, disseminating a
Christian worldview.

scholars—accepted what the Bible told them as the literal truth. They believed that their task was not simply to record or measure. Rather, it was to fit the world and what they thought they knew of it into the prevailing philosophical and religious viewpoints of the time. Medieval world maps, therefore, are as much concerned with spiritual development—from the Creation and the giving of the law to the coming of Christ and the Last Judgement—as they are works of geography, if, indeed, they can be described as the latter at all. The maps faithfully reflect the prevailing belief in the existence of divine order and a divine plan. Whether consciously or unconsciously, they served to validate the authority of the Holy Scriptures and thus supported the tenets and beliefs of the Christian faith.

Medieval mapmakers were by no means ignorant: it is highly likely that the scholars of the time salvaged much more from the ruins of the ancient world than had previously been thought to be the case. Nor were these mapmakers lacking in intelligence: their mindsets were simply different from ours. This is why the world maps they created prioritized religious over purely geographical content. The dominant idea was to give viewers a notion of the geography of the Christian world and that of its neighbors, relating the places shown on the maps to Biblical writings wherever possible. The result was a kind of geographical encyclopedia: an inspirational guide to the knowledge of the times.

This knowledge was greater than is sometimes thought. Medieval mapmakers obviously knew about Europe, though not so much about its northern extremities. They also knew a reasonable amount about North Africa and the Middle and Near East. However, they knew little or nothing about much of Asia—China, thanks to Marco Polo and other travelers, being the exception—the greater part of Africa, what lay beyond the shores of the Atlantic, and the southern hemisphere as a whole. Inevitably, beyond these limits geography became more speculation than science, until the Portuguese voyages of exploration and discovery down the African coast and beyond the equator in the 15th century really changed people's view of the world.

EARLY MAPPAE MUNDI

Probably the best-known medieval maps are *mappae mundi*, over 600 of which have survived. Many of them were very small, being created to illustrate books, but others were extremely large and obviously intended for public display. The reasons for the maps' creation varied. Secular rulers saw them as visual attestations of their power, prestige, and intellectual prowess, while churchmen probably employed them as educational tools to help them to instruct the faithful. They were not—and were never intended by their makers to be—geographical guides.

Nor were these world maps intended to stand alone. All had accompanying texts in the form of detailed annotations and commentaries. This is not surprising, since most other medieval maps were created to illustrate manuscripts, not necessarily geographies but, more often than not, works such as Bibles, calendars, scientific treatises, chronicles, and histories. By examining *mappae mundi* in context, it is possible to detect how medieval thought determined the forms and content of maps and, in turn, what part maps played in shaping it.

T-O maps

The majority of *mappae mundi* were designed as what are termed T-O maps. This is because they look very much like a T within a circle or an oval. In some instances, the result can look more like a diagram than what we expect a map to look like today. Some are centered on Jerusalem; other centers include Mount Sinai and, naturally enough, Rome.

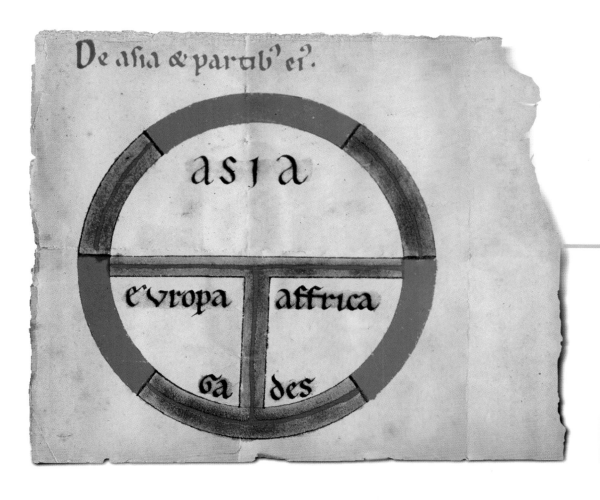

13 *Mappa Mundi,* Spain, 7th century This diagrammatical T-O world map was devised to illustrate the *Etymologiae,* an encyclopedia of knowledge compiled by St. Isidore, archbishop of Seville from 600 until 636. The map's actual author is unknown. It followed the ancient Greeks in dividing the world into the three continents of Asia, Africa, and Europe, surrounded by an encircling ocean. It was frequently copied during the Middle Ages; in 1472, it became the first map ever to be printed.

The T divided the Earth's landmass into the three known continents—Asia, Africa, and Europe—often with Asia at the top, Africa to the right, and Europe to the left. Asia is always the largest region because, according to Isidore of Seville, the 7th-century bishop and scholar, St. Augustine had said that it was "the most blessed." The upright of the T stood for the Mediterranean, separating Africa from Europe, while the right side of the crossbar represented the Red Sea or the River Nile, dividing Africa from Asia. The left side stood for the Black Sea, the Sea of Azov, and the River Don, which separated Asia from Europe. One theory is that the T symbolized the tau cross (named after the Greek letter it resembles), which features in both the Old Testament and in early Christianity. The O is similarly a convention to represent the encircling ocean.

Although some world maps were orientated toward the north, medieval mapmakers chose east as the orientation for most of them. Not only is sunrise in the east, but the mapmakers also believed that the Garden of Eden was located in the easternmost part of the world. Consequently, it often became the convention to show Eden at the eastern extremity of such maps.

The Beatus and Anglo-Saxon world maps

The earliest medieval world maps appear to have drawn heavily on Roman models and are more formal in style than later examples. Dating from the 8th century, the Albi map is one of the oldest *mappae mundi* still in existence. Its geographical content is simple, with the habitable world depicted as an oblong, rounded at the corners and surrounded by ocean. The emphasis is on the

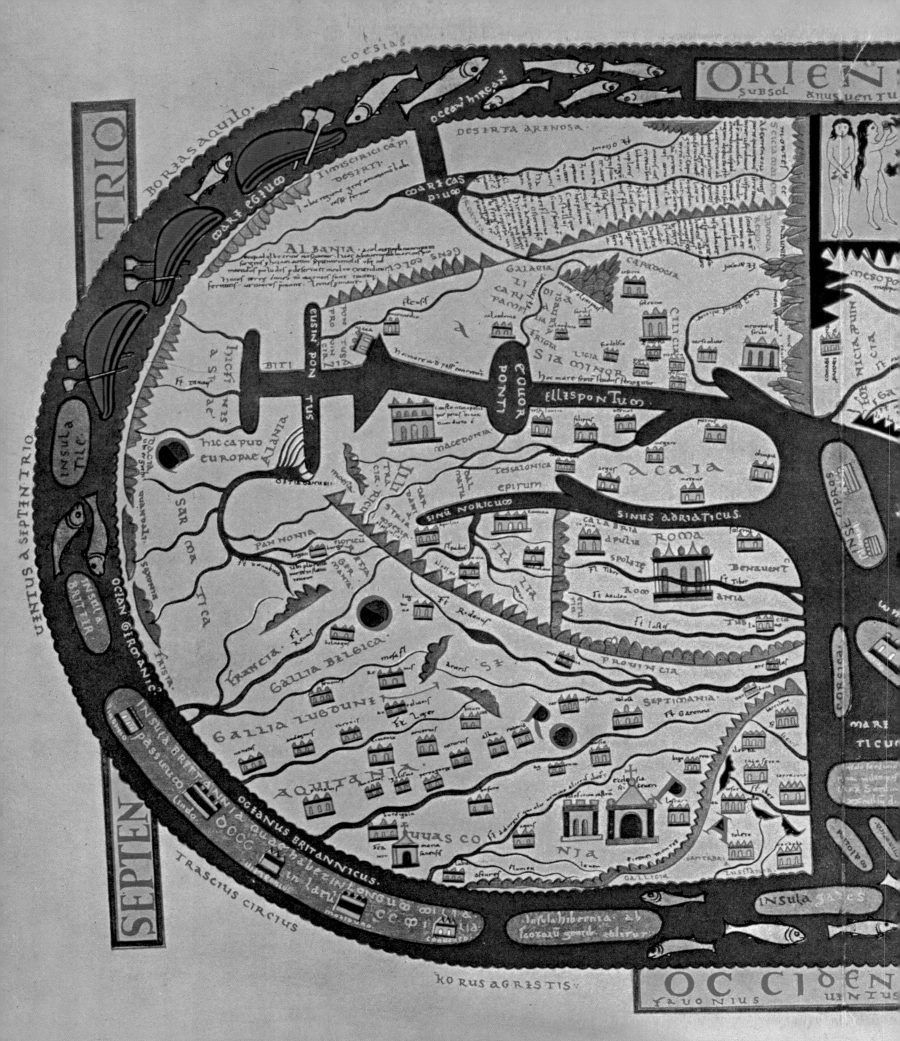

14

***Mappa Mundi*,**
Beatus of
Liébana, Spain,
c. 776

Beatus of Liébana created his world
map around 776; this is an 11th-
century copy executed by a monk at
the monastery of St. Sever in France.
Unlike most other medieval world
maps, it features a fourth continent,
represented as a strip of land
running along the southernmost
edge of the Earth beyond the
equatorial ocean, on the right-hand
edge of the map. According to
Beatus, this was where "the
Antipodeans are fabulously said
to dwell."

15

Anglo-Saxon *Mappa Mundi*, England, 10th or early 11th century

This map is different from other world maps of medieval times. It is square rather than round, and is surprisingly accurate, despite not being based on a survey. It seems likely that it derives from a map from Roman times, which may have been based on geographical coordinates, the knowledge of which was lost, or certainly not understood or used by, the medieval mapmakers.

Mediterranean and its surrounding lands. There are numerous misrepresentations: Judea, for instance, appears south of the Mediterranean, Crete to the north of Cyprus, and Sardinia to the north of Corsica.

Dating from several centuries later, the Beatus world map resembles the Albi map in its structure, with the Mediterranean at its center, and the Aegean, Adriatic, and other seas branching from it (see pp.34–5). The original map was made to accompany the *Commentary on the Apocalypse of St. John*, produced by Beatus of Liébana, a Spanish Benedictine monk, in 776. Fourteen copies of the original, created for Beatus by an anonymous mapmaker, survive, dating from the 10th to the 13th centuries. The map shows the world in which the Twelve Apostles were sent to proselytise. The Mediterranean is in the middle of the map with the River Nile to the right and the Danube and Don forming a triangle on the left. Jerusalem is a three-turreted spire at the top of the Mediterranean, with Adam, Eve, and the serpent depicted to the right of it. A thick red strip of sea to the right of that separates a fourth continent from the rest of the world. The accompanying annotation states that the land beyond this sea was unknown to humans owing to the great heat of the sun. It is arguably the antipodean continent south of the equator.

The Anglo-Saxon or Cottonian world map, which was probably drawn in Canterbury in the 10th or early 11th century, shows considerable Roman influence, notably in the amount of detail it includes and in the very recognizable coastline it gives to Britain. It is thought probable it was a much debased copy of a map that was first drawn in the Roman period. The anonymous cartographer is likely to have been an Irish scholar-monk who worked in the household of Archbishop Sigeric of Canterbury, since the map draws on the itinerary the archbishop composed covering Europe from Rome to the English Channel.

Itineraries

Itineraries, or list maps as they are sometimes termed, superficially looked like T-O maps, although, in fact, they were not maps at all, at least as we understand the term, as they contained no element of plan. Instead, they listed in written form the characteristics of different lands, or the different places to be found in them. There was little or no attempt to establish a geographical relationship between the entries. A classic example of this kind of map is the one that was possibly compiled by a cleric in Ravenna shortly after 700 to illustrate a text he had written for the benefit of one of his fellow clerics.

The Ravenna Cosmographer, as he is known, started by outlining the geography of the world, making frequent reference to the Bible and to Christian cosmology. The nations of the world were then listed in two distinct groups, a fact that suggests that the map or maps that are thought to have been created to accompany the text were either elliptical or circular. One grouping extended from India to southern Ireland via Africa, and the other covered the area stretching from central Ireland via what the Cosmographer called Scanza Island (modern-day Scandinavia) and Scythia to India.

There are more than 5,000 place names listed with separate sections for rivers and islands. There are also lists of tribes corresponding to orientation. It is a monumental work, so it is not surprising that there are numerous errors and corruptions in the compilation, either through carelessness on the part of the Ravenna Cosmographer or created by later copyists. Quite frequently, for instance, river names are wrongly entered as if they were places. Often, the lists appear as though they are following a logical topographical sequence, only for names from a totally different part of the region to be suddenly interposed.

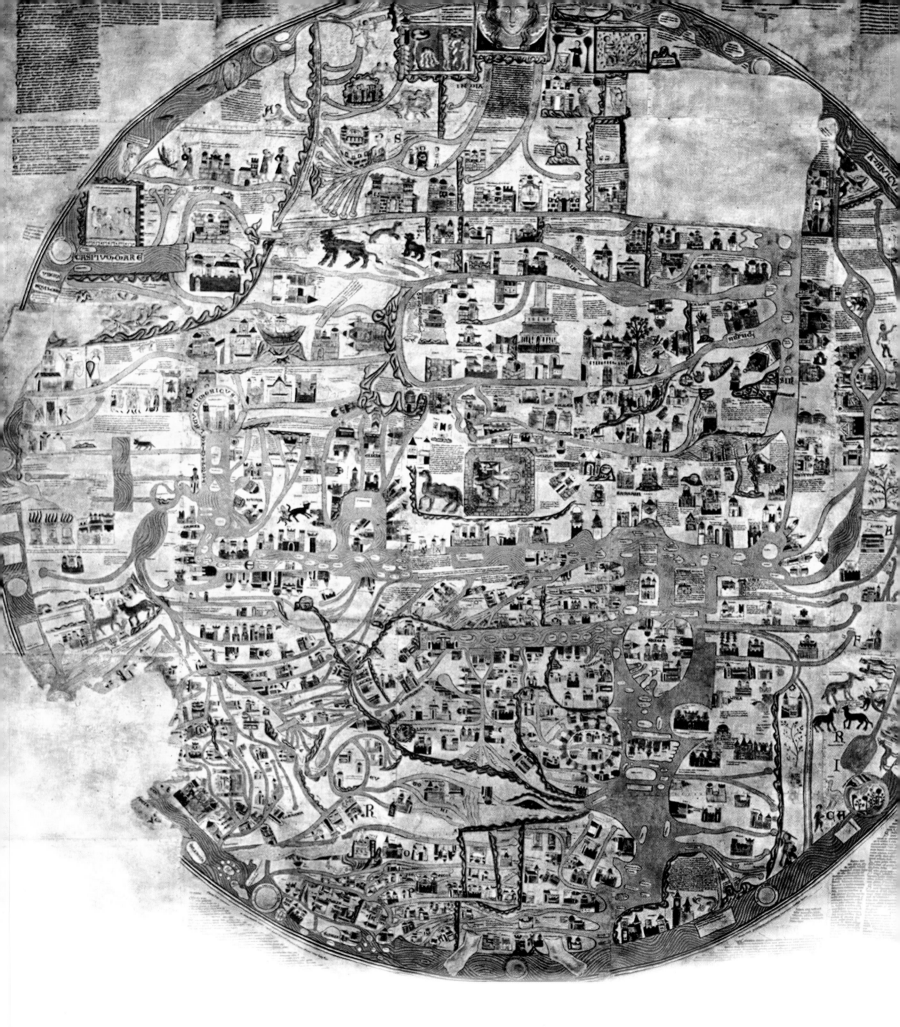

WORKS OF ART

As the medieval period progressed, *mappae mundi* were rendered in greater and greater fantastical style. Decorative elements became more dramatic, with angels, legendary peoples, and mythical creatures, such as bonacons and unicorns, appearing. Prime examples of these great creations are the Ebstorf and Hereford maps, both dating from the 13th century. They reveal the wonderful religious cosmology that evolved during the Middle Ages.

Although the individual shapes of landmasses began to be better defined and labeling became more geographically related, the cartographers' chief aim remained the representation of the world as a Christian allegory. Such maps also acted as a form of visual world history, illustrating what it was like to live in different parts of the world, and revealed what was thought to have happened in them in the past, with visual coverage of events such as the crossing of the Red Sea or the building of the Tower of Babel. It was common for mapmakers to overlay civilizations, events, and locations from different time periods in order to create a composite view.

The Ebstorf map

Though not the largest world map known—the mural map at Chalivoy-Milon, near Bourges, which was destroyed during the restoration of the church that housed it in 1885, was 19 feet 6 inches (6 m) across—the Ebstorf map was nevertheless extremely impressive. Covering an amazing 30 sheets of vellum and more than 9 feet (3 m) wide, it got its name from the village of Ebstorf, near Ülzen, on the Lüneburg Heath in northern Germany.

The map was created towards the end of the 13th century, but who actually created it is unknown. The once widely held belief that it was drawn by the English cleric Gervase of Tilbury has been discredited. Some cartographic historians think that it might have been adapted from a map made slightly earlier for Otto the Child, Duke of Brunswick. What we do know is that the map remained housed in obscurity in a Benedictine nunnery in Ebstorf until its rediscovery in 1830 and subsequent removal, first to Hanover and then to Berlin, in 1891, where it was restored before being sent back to Hanover again.

The photographs taken at the time of the restoration are the only record of what the map looked like, since it was destroyed in an Allied bombing raid in 1943. This was a cultural tragedy, made the more poignant by the fact that it had been agreed that the map should be evacuated from Hanover to somewhere safer, but formal permission to do so had not been received from the authorities in Berlin before the bombers struck.

Whoever the mapmaker was, he seems to have had two things in mind when he created the map. Although his primary intent was to

16

Ebstorf *Mappa Mundi*, Germany, c. 1300
This is a reconstruction of one of the largest world maps ever created. The original was destroyed in an air raid during World War II. Designed to instruct the faithful about significant events in Christian history, its religious purpose is clear. The world is depicted as the body of Christ, his arms embracing it and its peoples, while important places and incidents from the Bible, such as Noah's Flood and the parting of the Red Sea, are prominently depicted.

illustrate events in the Christian story—the overall focus is on salvation—he might have intended his map to be of some practical use, which would have been unusual. According to one of its inscriptions:

> " *It can be seen that [this map] is of no small utility to its readers, giving directions for travelers, and the things on the way that most pleasantly delight the eye.* "

The sources that were drawn on for the purposes of compilation were many and varied. Some were classical, such as the works of the Elder Pliny, while others, such as the *Imago Mundi* (an encyclopedia of popular cosmology and geography combined with a chronicle of world history) by Honorius Augustodunensis (d. 1151), the writings of Johannes of Würzburg (c. 1165) on Palestine, and those of Adam of Bremen (c. 1129) on northwest Europe, were medieval. The mapmaker naturally relied most heavily on scripture, the Fathers of the Church, and other Christian writings for knowledge and inspiration.

The content was superimposed over the figure of the crucified Christ, head at the top, feet at the bottom, and hands pointing north and south. Jerusalem was placed at the map's heart with Asia stretching east from it. Here, among other features, the mapmaker located the Garden of Eden, sheltered behind impenetrable mountain ranges, below which the River Ganges and its 12 tributaries flowed through a tropical landscape. On a lower level to the left was China, similarly hemmed in by mountains. In northern Asia, the main feature was a promontory that projected as a rectangle into the cosmic ocean. This was the home of the dreaded man-eating giants Gog and Magog; the castellated lines mark the massive walls that Alexander the Great was reputed to have built there to protect humanity from their incursions. Slightly to the west, two doughty warrior-queens guard the land of the Amazons, while still further to the west, flaming altars mark the northern extremities of the world. Due south, the city of Colchis stands on the Black Sea, the fabled Golden Fleece that Jason and the Argonauts set sail in search of hanging from a tower, while above and to the right is Mt. Ararat, with Noah's Ark stranded upon it.

Africa, by contrast, was mapped more sketchily. It was depicted as little more than a segment of a circle, its north and west coasts running in almost a straight line from the Indian Ocean to the Atlantic. Its south and east coasts describe a shallow curve. The principal feature is the River Nile, which the mapmaker seems to have believed had its source in a lake in modern-day Morocco. The river's initial course is west to east until, as it nears the eastern tip of the continent, it disappears into the sand. It emerges flowing in the opposite direction through Egypt, having first skirted a land inhabited, so it seems, by a race of dwarves who rode on crocodiles.

Both Asia and Africa were littered with pictures of weird and wonderful peoples, animals (there are about 60 depicted on the map in total), and other strange mythical beings and creatures. Though the mapmaker obviously knew more about Europe, it was given the same stylized treatment, with little attempt to render coastlines or other geographical details accurately. For these reasons, many later historians condemned the map as worthless, but this was to misunderstand completely the times in which the map was created and the society that created it. In many ways, it is a realistic depiction of the world as the majority of medieval Europeans saw it.

The Hereford map

With the loss of both the Chalivoy-Milon mural map and the destruction of the Ebstorf map, the Hereford *mappa mundi* took their place as the largest surviving medieval world map. It is certainly

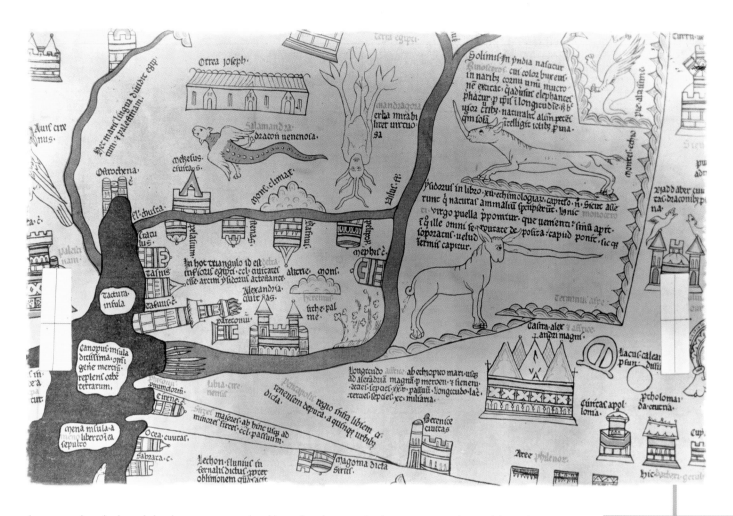

the most detailed and the best preserved, although when and why it was made and by whom are matters of debate. It now seems most likely that it was created in Hereford in around 1300, possibly in the hope of furthering the call for the canonization of Thomas Cantilupe, bishop of Hereford until his death in 1283, and the career of Bishop Swinfield, Cantilupe's successor. The likelihood is that the map is a copy of an earlier map that Richard de Bello, prebendary of Sleaford, had designed at Lincoln some time before his death in 1278. No trace of this earlier map has survived, although it is thought that Richard might have consulted the instructions for drawing world maps contained in the *Expositio Mappae Mundi*, a treatise possibly compiled by Roger of Howden, a Yorkshire cleric who had accompanied Richard I on the Third Crusade, in order to produce it or supervise its production.

Whatever its origins, the map was and is tremendously impressive. Over 5 feet (1.65 m) high and almost 4 feet 6 inches (1.35 m) wide, it was drawn on a single sheet of vellum. It is thought that it might have been intended as a decorative background for an altarpiece, or for educational purposes to aid in the teaching of the Christian faith. The map itself is contained in a circle 52 inches (1.32 m) in diameter. According to an inscription on the map, the information on it was based on the works of a pupil of St. Augustine called Orosius and the description of the world he gave in his *Ormesta*, which he wrote in the 5th century. As is commonly the case with similar medieval maps, the known world is represented as a flat disc encircled by sea with the east at the top. Here Christ is portrayed presiding over the Last Judgment, angels ranged on either side of him, the saved to the left and the damned to the right. The Virgin Mary pleads on behalf of humanity.

17

Hereford *Mappa Mundi*, England, c. 1290–1300

An undisputed masterpiece of medieval cartography, the map includes more than 1,000 inscriptions and as many pictorial representations of places, peoples, and animals believed to be found in various parts of the Earth. The result is an encyclopedia of world knowledge, as it existed at the time. This section shows the Nile Delta. The lighthouse at Alexandria, a rhinoceros, and a unicorn can be seen.

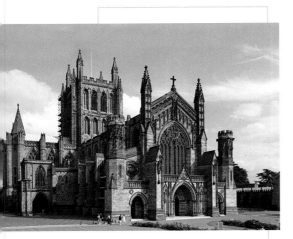

Hereford cathedral, England
Much of Hereford cathedral, including the nave and central and western towers, dates from the 12th century. It is believed that the Hereford *mappa mundi* was created to further the call for canonization of the bishop of Hereford, Thomas Cantilupe, in the late 13th century. He was canonized in 1320.

Christ's arms are outstretched as if to encompass the whole of the world. The Garden of Eden lies just inside the disc, directly under the figure of Christ. A stone wall and a ring of fire surround it, with Adam and Eve standing outside its locked gates. Balancing Eden, located at the eastern limits of the map, the Pillars of Hercules stand at the western limits.

The map is packed with inscriptions and legends, many of which are the names of towns and cities; others include rivers, mountains, islands, and seas. Hereford itself appears alongside the River Wye, its location marked by a drawing of what may be its cathedral, while there is also a depiction of Lincoln on its hill. Some of the imagery is extremely up-to-date: some of the Welsh and English castles relatively recently constructed for Edward I are depicted.

Other inscriptions provide detailed cosmological, ethnographical, historical, theological, and zoological information, frequently with an appropriate illustration next to them. In total, there are 500 or so drawings of cities and towns; 15 re-creations of notable Biblical events; 33 depictions of plants, animals, birds, and strange creatures; 32 portrayals of the peoples of the world; and 8 pictures from classical mythology. There are numerous tableaux, examples of which include a mother tigress foiled by a hunter with a mirror, and a griffin fighting with three men over a hoard of emeralds. The real and imaginary beasts depicted on the map include the dog-headed Cynocephali, the shadow-footed Sciapods, a manticore, a parrot, a crocodile, and a unicorn.

Several craftsmen created the map, probably working in stages. The three outer circles were most likely laid out first with a compass before the defining geographical features—the coastlines of the three continents and most or all of the map's 105 islands—were outlined. Then came the pictures, which were probably the work of a group of artists. These are likely to have been drawn all at one time, almost certainly before the addition of any written text.

Much of the writing on the map is in black ink, with red and gold leaf used for emphasis. Rivers and seas are colored blue and green, although the Red Sea, naturally enough, was depicted as red. The colors were made from vegetable dyes, which have faded over time so that what we see today is far less vivid than would have appeared to medieval viewers. Scalloped designs indicated mountain ranges, and walls and towers depicted towns. The language is mainly Latin, although Anglo-Norman French is used for special entries.

All the legends on the map were the work of a single scribe, although the task of actually naming the continents was given to another. This unknown calligrapher was responsible for the map's greatest error: he inadvertently transposed the labels for Europe and Africa.

ZONAL MAPS

Another form of map fascinated many medieval cartographers: so-called zonal maps. In these, according to medieval theory, a spherical world was divided into three uninhabitable zones— frozen ones at the two poles and a torrid region at the equator—with two mutually inaccessible but habitable temperate zones sandwiched between them. The zones roughly correspond to what we would term climatic belts.

The idea was not originally a medieval one. Rather, it was the Classical Greeks who first thought of it, possibly as early as the 5th century B.C.E. Some experts believe that the man who first put forward the notion was a scholar called Parmenides of Elea (born c. 515 B.C.E.), but others give the credit to Crates of Mallos, a scientist who lived a couple of centuries later. The Greeks passed the theory on to the Romans, when Ambrosius Aurelius Theodosius Macrobius, a Neoplatonist philosopher and grammarian, presented it in finite form in his *Commentary on the Dream of Scipio According to Cicero*, which he wrote in around C.E. 400. His basic belief

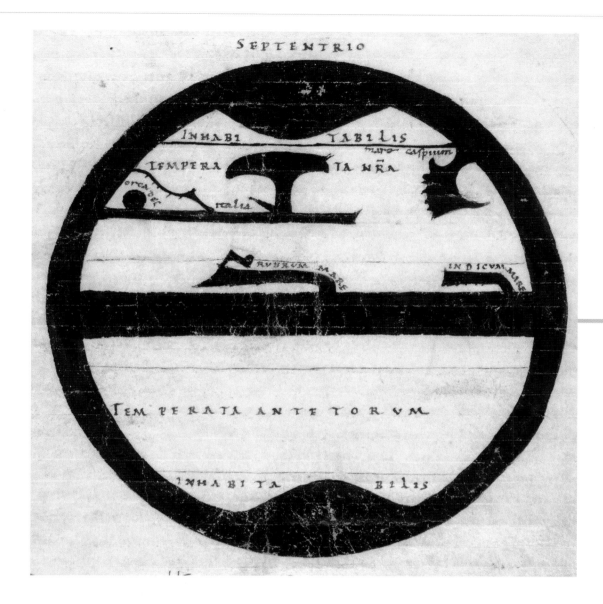

SEPTENTRIO

INHABI   TABILIS

mare caspium

TEMPERA   TA N̄RA

orcade

rralis

RUBRUM MARE

INDICUM MARE

TEMPERATA ANTETORUM

INHABITA   BILIS

18

Zonal World Map, Spain, 9th century Dividing the known world into zones according to climate was a Greek idea subsequently refined and developed by the late Roman thinker Macrobius in the early 5th century C.E. His world map was frequently copied, this copy dating from the 9th century, and subsequently became a basic reference for medieval scientists. The world was divided into five zones, two of them freezing, two temperate, and one very hot around the equator. From left to right beneath the northern, freezing zone, which is uninhabitable ("inhabitabilis") is the Caspian Sea (with the Indian Ocean beneath), the Black Sea, Greece, Italy ("Italia"), and a bump representing Spain. Beneath these is a horizontal strip representing the Mediterranean and under it an L-shaped Red Sea. At the left, forming a rough triangle with a circle in the middle, are the Orkney Islands ("Orcades").

was that the Earth was round, stationary, and at the center of the universe. The inhabited world, he thought, was shaped like a lozenge—with Italy firmly at its center—and surrounded by ocean, as were the other three landmasses which he thought existed in the remaining quarters of the world.

Macrobius' texts survived the fall of the Roman Empire and the Dark Ages that followed to become a lynchpin of European medieval thought. The maps that were based on his writings were first drawn in Spain in the 8th and 9th centuries and later reproduced in the works of Lambert of St. Omer, William of Conches, Honorius Augustodunensis, and other notable medieval scholars. They showed the inhabitable world of the northern hemisphere and the uninhabited one of the southern marked with climatic zones, which Macrobius had derived from Ptolemy's climata. The maps were orientated with north at the top. As the space allocated to the northern temperate zone was necessarily relatively small, only a few place names—the Orkney Islands, off the coast of Scotland; the Caspian Sea; the Indian Ocean; and the Red Sea—were shown, selected because they marked the extremities of the known world. The notion of a balancing landmass lying somewhere between the equator and the southern polar region was purely speculative, as Macrobius himself had recognized, as was the question of whether it was or was not inhabited or habitable. Medieval scholars followed him in arguing that, without the

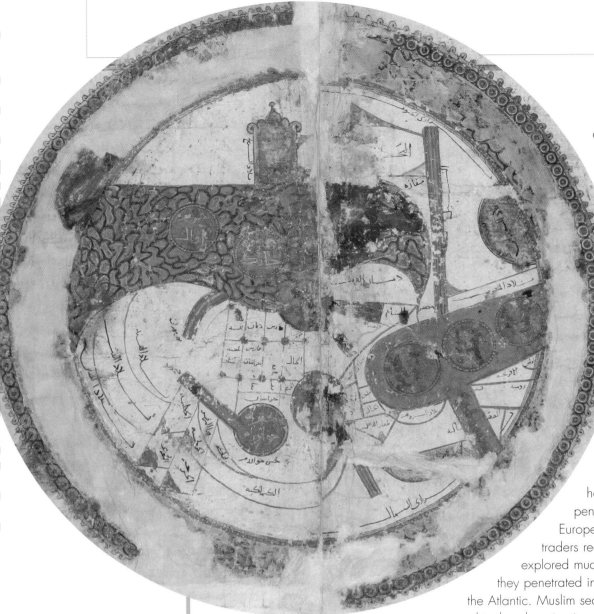

existence of what they called the Antipodes, there was no possibility of maintaining the overall symmetry that was a vital constituent of God's divine plan at the time of the Creation.

MAPPING IN THE ISLAMIC WORLD

While knowledge of Ptolemy's work was lost to Western Europe for many centuries following the fall of the Roman Empire, the Islamic world was more fortunate. In the 9th century, Ptolemy's *Geography* was translated into Arabic. Muslim mapmakers were also to benefit from the discoveries of Arab explorers. Conquest had secured Islam control of the Middle and Near East and North Africa; it had made the Muslims masters of the Iberian peninsula and taken them deep into Western Europe. Between the 7th and 9th centuries, Muslim traders reached China, and they subsequently explored much of the east coast of Africa. Northward, they penetrated into Russia, while westward they sailed into the Atlantic. Muslim seafarers eventually also had new or improved geographical and navigational instruments to help them, notably the astrolabe and the compass. Knowledge of the latter reached them from China.

Among the first notable Arab cartographers was Al-Khwarizmi, who was born in Baghdad in around 780 and died there in around 850. Though first and foremost a mathematician—he is credited with establishing the basics of algebra—he was also interested in mapping and mapmaking. He made a compilation of 2,402 localities, probably using Ptolemy's *Geography* as his base, but improving on it, particularly in his location of cities in Asia and Africa. Like a similar map that had been prepared by a group of anonymous Baghdad cartographers for Caliph al-Mamun in about 820, it portrayed "the universe with spheres, the stars, land and the seas, inhabited and barren, settlements of people, and cities."

Another important figure was al-Istakhri (d. 951), an influential member of the Balkhi School of Islamic cartographers (the school was named after its founder, geographer Abu Zayd Ahmad ibn Salh al-Balkhi). His original maps, like his first major treatise, *Suwar al-aqalim* (Picture of the Climate), have not survived, but his cartographic commentary was expanded by ibn-Hawqal, a Sicily-based geographer who died not long after 973, into a work called *Kitab surat al-ard* (Picture of the Earth), which is known in three versions dating from the mid- to late 10th century. As well as a world map, this featured maps of the Caspian Sea, the Mediterranean, and the Indian Ocean, plus what can best be described as provincial or regional maps. These last are little more than maps of camel routes, marked with staging areas, villages, and oases.

The world map is of great interest. Orientated with south at the top, centered on Mecca, and drawing heavily on Islamic reference as a reaction to the Ptolemaic and Western influences that were now permeating much of Muslim cartography, it is focused on the Baghdad Caliphate and the vast empire it controlled. The inhabitable world is portrayed as a circle surrounded by mountains and an enclosed sea. The map is split more or less in half by the Indian Ocean and the Mediterranean, with Africa at the top of the map, bisected by the River Nile. Europe lies to the lower right, with Greece and Italy extending upward into the Mediterranean. The map is highly stylized, reflecting the aesthetic values of Islamic society and its political and religious views.

The Book of Roger

The most celebrated Islamic geographer of the medieval period was undoubtedly al-Idrisi (1099–1166/80). A Berber born in Ceuta in present-day Spanish Morocco (he is believed to have been a descendant of the Prophet Muhammad), al-Idrisi studied at Cordoba before embarking on his travels, which took him through Spain and North Africa before he ventured further afield to England, France, and Asia Minor. Eventually, he settled in Sicily, where he compiled a systematic geography of the world. He called it *Nuzhat al-Mushtaq fikhtiraq al-afaq* (A Guide to Pleasant Journeys into Faraway Lands), but it is more commonly known as the *Book of Roger* after Roger II (1097–1159), the Norman ruler of Sicily who was al-Idrisi's patron. The king was a fervent devotee of geography, much of his spare time being taken up with the collecting of Arabic geographical treatises and in questioning travelers visiting his court in Palermo about their experiences in distant lands.

Roger commissioned al-Idrisi to create his monumental work in 1139 (see p.46). Completed 15 years later, it was a detailed world survey, consisting of a small, disc-shaped world map and 70 rectangular sectional maps complete with an accompanying narrative in both Arabic and Latin. Following Ptolemy, the inhabited world was split into seven zones, each stretching horizontally northward from the equator, while ten sections extended vertically from the Canary Islands eastward. To help in the compilation, king and mapmaker selected what the chroniclers termed "certain intelligent men," and despatched them, together with draftsmen, to bring back detailed records of what they had found out on their travels. The result was a cartographic triumph, even though some scholars have quibbled about the fact that East Africa and India were mapped surprisingly sketchily since both regions were known to Muslim merchants of the time. The information the maps carried was first engraved on a silver planisphere (a representation of a sphere on a plane surface), which at the time was considered one of the wonders of the world. Unfortunately, the planisphere was destroyed by a rioting mob in 1166, but not before the information it carried had been transcribed onto parchment.

PORTOLAN CHARTS

In the late 13th century, medieval European mapmaking underwent something of a revolution with the arrival of portolan charts. Unlike the great world maps, which are probably best described as cosmological rather than geographical, these new maps were eminently practical, devised in particular circumstances to solve particular problems. Portolan charts were no-nonsense working tools, designed to assist sailors and incorporating the wisdoms and knowledge that generations of seafarers had accumulated. Also noteworthy is the fact that the charts were complete in themselves, rather than having to be studied in conjunction with an accompanying text or

Magnetic compass, Arab, 9th or 10th century
The compass is the oldest instrument of navigation. Use of it spread from China through Arab traders to Europe, and by the late 12th century simple magnetic compasses were being used for navigation in the Mediterranean.

20 World Map, al-Idrisi, Sicily, 1154

The Berber mapmaker al-Idrisi was one of the leading geographers of medieval Europe. This world map, which he created for his patron Roger II of Sicily, was part of a comprehensive atlas that included 70 sectional maps. The map is probably the most accurate depiction of Europe, North Africa, and western Asia to be produced in medieval times. It is orientated with south at the top, with the Mediterranean two-thirds of the way down. The craftsmanship is exquisite, especially in the use of colors—purple and ocher for mountains, green for rivers, and blue for seas—and elaborate, beautifully worked symbols.

commentary. As such, they represented a significant evolutionary development in the story of cartography.

Exactly how, why, when, and where the first of these sea charts came into existence remains one of cartography's greatest mysteries. It is almost as if they burst upon the medieval world fully formed, but this seems highly unlikely. In part, their creation might have been sparked by the ever increasing importance seaborne trade was assuming in the Mediterranean, as the rise of city-states such as Venice to commercial and political prominence aptly demonstrated. The power, prestige, and wealth of the Venetian Republic were built on its trading fleets.

What we do know is that the charts' direct precursors were the so-called *portalani*, elaborate lists of ports and written sailing directions that had been used by sailors in the Mediterranean for many centuries. It would have been impossible for the new portolan charts to be devised before knowledge of the magnetic compass reached Europe via the Muslim world in the late 12th century. The earliest record that exists of any form of chart being used on board a ship dates from 1270, when, while en route to launch the Eighth Crusade, Louis XI of France was forced to take refuge from a storm in Cagliari Bay on the coast of Sardinia. According to the chroniclers, when the irate king demanded to know exactly where his ship was, the sailors brought a map to show him their position.

The Carte Pisane

The oldest surviving portolan chart dates from around 1290. It shows the Mediterranean and the Black Sea. It was christened the Carte Pisane because a family who lived in Pisa originally owned it, but exactly who compiled it is an unsolved riddle. Most scholars believe that it was drawn up in the wealthy trading port of Genoa, which was, along with Venice, one of the first great centers for such charts' production. It is likely that the chart was an amalgam of information drawn from earlier charts of limited stretches of coastline sailed and charted by local fishermen and the captains of small coastal traders. The chart is far too accurate to be the work of only one man, as no cartographer working alone could have hoped to cover so large an area in such intricate detail (see p.48).

The Carte Pisane contained all the elements that featured in subsequent portolan charts, regardless of whether they were produced in Italy or by the later, rival school of Catalan chart-makers. The main difference between the two schools is that Italian charts lacked interior detail of landmasses apart from the flags of rulers, while the Catalan ones were far more decorative, thus bearing a passing resemblance to the *mappae mundi* being produced in northern Europe. Either way, the instantly recognizable feature is the web of so-called rhumb lines crisscrossing such charts as they radiate from a central compass star through sixteen intersecting ones. Rhumb lines and stars together provided a ship's navigator with all he needed to lay an accurate course. Later, as a further navigational aid, the lines were color-coded according to prevailing wind directions. The eight principal winds were drawn in black, the half winds in green, and the quarter winds in red.

Such charts also featured a scale, whose divisions notionally corresponded to the dimensions of the quadrilaterals created by the intermeshing of the rhumb lines. The scale was drawn on a ribbon-like strip that to modern eyes looks almost like a tape measure. The depiction of the coastlines was remarkable for its detail and, on the whole, its accuracy, although the chart-makers obviously knew more about some parts of the area they were depicting than others. Because of the navigational assistance they provided, the shapes and sizes of headlands, capes, harbors, and river estuaries were deliberately exaggerated, while, to avoid obscuring the coastline, place names were written at right angles to it on the land. Islands and coastal cliffs were often drawn in perspective.

Important ports were marked in red, while dots and crosses signified potential hazards, such as sandbanks and rocks. Details of the interior, such as mountain ranges, cities, and roads, were either glossed over or omitted altogether. The language used for the chart legends was almost always Latin, although a few charts employed either Catalan or Italian dialects.

From chart to atlas

Having successfully created individual portolan charts, it was natural enough for cartographers to start producing portolan atlases, the earliest of which date from the 14th century. Such atlases contained a world map (usually, this was drawn as an oval); local charts showing particular stretches of coastline; separate charts of the Adriatic, the Aegean, and sometimes the Caspian

21 Carte Pisane, Italy, c. 1290 Known as the Carte Pisane because it once belonged to a wealthy merchant family from Pisa, this is the earliest surviving example of the revolutionary navigational charts that were to transform medieval cartography. Such charts were intended to be practical tools for sailors in the great age of Mediterranean trade, so they concentrated on accurate depiction of coasts, landmarks, and harbors. The lines on the chart—each radiating from a compass point—were further navigational aids.

Seas; and a chart of the entire Mediterranean. In addition, such atlases frequently contained supplementary sailing directions, astronomical and astrological calendars, and sometimes tables of the phases of the Moon. Sometimes, too, other items of useful practical information were appended. When Andrea Bianco, the commander of a Venetian galley, compiled his own portolan atlas in 1436, he provided just such an appendix. It contained instructions as to how best to calculate dead reckoning when tacking or running before the wind.

Undeniably, the finest of these atlases is the so-called Catalan Atlas (see p.30). Produced in 1375 by Abraham Cresques, a Jewish book illuminator and, according to his description of himself, "master of maps of the world and compasses," it is a cartographic marvel that has a pivotal place in the story of the development of Western cartography. Cresques was the most renowned member of the celebrated Majorcan cartographic school that came to dominate 14th-century European mapmaking, so it was naturally to him that King Pedro IV of Aragon turned when looking for a cartographer with the ability to create a suitable gift to please Charles V of France, who had asked the Aragonese king to create a map of the world for him.

The atlas that resulted is unique. Not only was it a true reflection of the state of geographic knowledge of the time, but it also provides a comprehensive insight into the worldview held by Europeans of the late 14th century. The maps it contained were the most accurate of the day.

The atlas was originally created on six sheets of vellum, which were then stretched out flat on wooden panels. It is thought this was a device to suit the needs of Charles and his advisers, since they could consult the map flat when in council. Later, the sheets were folded in half so that the atlas could be bound together as a book. The opening four sheets featured elaborate discourses on cosmography, astronomy—this part of the atlas includes the earliest surviving set of lunar tide tables—and astrology. This last section features a magnificent astrological wheel that includes depictions of the zodiac, what medieval thinkers believed were the elements constituting the Earth, the planetary deities, and a calendar wheel. It was the dating on this wheel that enabled later scholars to date the atlas precisely to 1375.

The remaining sheets were devoted to an elaborate world map. This is orientated with south at the top. In compiling this map, Cresques was influenced by the two different Western mapmaking traditions: portolan charts and *mappae mundi*. For Europe, which is by far the most detailed portion of the map, he undoubtedly had access to many of the portolan charts produced by the Catalan school of chart-makers, while he was also probably familiar with similar charts drawn up by Genoese and Venetian chart-makers. As a result, the European seacoast is outlined extremely accurately. The whole western section is crisscrossed with rhumb lines, while in mid-Atlantic Cresques positioned a large, colorful compass rose—the first known to appear on any portolan chart—with an eight-pointed pole star indicating the north.

Following established portolan custom, place names were written at right angles to the coastline, while flags were used to show at a glance what the dominant political power was in any area. Away from the sea, mountain ranges and rivers are clearly marked, while church spires and crosses denote major cities (outside Europe, Muslim urban centers are identified with a dome). There are 620 place names mapped in Western Europe alone.

For regions further removed from Europe—particularly Asia—Cresques was less sure of his ground. This is why the Asian and African parts of his map look much more like traditional *mappae mundi*, with much extraneous illustration, as opposed to the European section, which was left relatively unadorned. Cresques made shrewd use of existing world maps and the writings of the great scholars of Classical times, such as the elder Pliny, as he embarked on his map compilations. He also had access to more contemporary references, such as the account Marco Polo (1254–1324) had written of his travels in China and Persia, while, for Africa, he referred to the chronicles of Ibn Battuta (1304–68/77), the great Arab explorer.

Thanks to Ibn Battuta and to Cresques' own travels, much of the information the atlas gives about North Africa is correct. The Atlas Mountains, for instance, although still depicted in traditional medieval form as a bird's leg and claw, are positioned correctly. The picture becomes cloudier the deeper the atlas penetrates into the continent and the further south it goes, with fewer and fewer land names. The illustrations that accompanied Cresques' depiction of Asia were particularly detailed. One of the most celebrated was of a camel caravan traveling along the Silk Road linking China with Persia. The accompanying text identifies places along the road reasonably accurately. This part of the atlas was the first map to achieve a recognizable presentation of Asia, with the Indian subcontinent, for instance, rendered as a peninsula for the first time.

Decorated compass rose, portolan chart, 16th century
An ornate compass rose at the center of 32 rhumb lines, indicating compass directions. The rhumb lines could be used by the chartmaker for plotting the coastlines and by the sailor for plotting their courses.

NATIONAL MAPS

Relatively few national maps were produced in medieval Europe. It was only in the 15th century, for instance, that maps were first employed to delineate frontiers. One of the earliest examples of such a map was one produced in 1444 by Henri Arnault de Zwollle, the trusted councillor of Philip the Good, ruler of the Duchy of Burgundy, to show territory whose rulership was in dispute between Burgundy and France. The Burgundian intention was to use the map to identify French enclaves and encroachments as a preliminary to forcing their removal.

There are a number of reasons for this seeming lack of interest in national maps, the most obvious being the prevailing political and social climate of the medieval period. Until perhaps the

22

**Britain,
Matthew Paris,
England,
c. 1250**

Matthew Paris, a Benedictine
monk at St. Albans' Abbey, drew
four maps of Britain in the mid-
13th century. This one features in
the *Abbreviatio Chronicorum*, an
abridgement of his three-volume
Chronica Majora. The boundary
between England and Scotland is
marked by Hadrian's Wall and the
Antonine Wall further to the
north, which was constructed
by Antonius Pius, Hadrian's
successor. Both are depicted
schematically as battlemented
features, although the Antonine
Wall was a ditch and turf-wall
fortification. As England's capital,
London has a particularly
elaborate towered and
battlemented frame around its
name. The spire to the north of
London marks the Abbey of St.
Albans. Paris's maps of Britain
stand out in the story of medieval
mapmaking, even though there is
nothing to suggest that any other
mapmaker of the times was
inspired by them or drew
material from them. Their
importance lies in the light that
they shed on the understanding
of cartographic concepts of one
of the most original thinkers of
the day.

15th century, people on the whole thought more in terms of localities, provinces, and regions than they did of nations. For instance, a person we would today define as a Frenchman was far more likely to think of himself as a Breton, a Norman, or a Burgundian.

There was little demand for national maps for another reason. Maps were not at this point thought of as a tool for travel. In fact, on the whole, the majority of medieval people were unlikely to have traveled more than 15 miles from their birthplaces during the whole of their lifetimes. If they did venture outside their own localities, medieval travelers seem to have used descriptions and gazetteers rather than maps.

Matthew Paris and his maps

Among the most notable maps of the medieval period were those produced by Matthew Paris, an English Benedictine monk born in around 1200 and based for much of his adult life at St. Albans Abbey in modern-day Hertfordshire. Fifteen of his maps have survived, including four versions of a map of Britain, the third of which was completed by Paris's friend John of Wallingford, the almoner of the abbey, adding 60-odd names of his own.

In his maps, Paris can be seen experimenting with cartographic concepts that were virtually unknown at that time. Very few maps existed—indeed the actual notion of a map was not generally understood. The maps of Britain are of particular interest. No similar maps of a single region of Europe of this time is known; the closest parallel is the depiction of Germany to be found in the Ebstorf world map. Even in the 14th century, the only comparable works are a map of Italy, drawn in Naples in about 1320, and the so-called Gough map of Britain, which dates from around 1360. The first of Paris's British maps appears at the beginning of the *Abbreviatio Chronicorum*, one of his abridgements of his three-volume chronicle, the *Chronica Majora*. The second comes at the start of the second volume of the *Chronica Majora*, the third was incorporated by John of Wallingford into a work of his own, and the fourth is at the beginning of a volume containing the *Historia Anglorum*, another of Paris's abridgements, and the final part of the *Chronica Majora* itself.

All four of the British maps probably took their coastal outlines from contemporary world maps—there are certainly striking points of resemblance between the Anglo-Saxon world map (see p.36) and Paris's creations—with additional information superimposed. To assist him in their compilation, Paris may have drawn upon information from contemporary writings and accounts from travelers. What he did not do was employ any form of survey, as this method of mapmaking would have been completely unknown to him and all other Western European mapmakers of the time. Like all his maps, the British ones were created solely in his study in St. Albans.

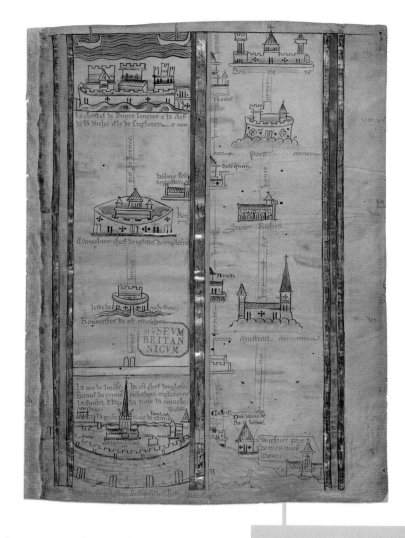

23 **London to Dover, Matthew Paris, England, c. 1250**
As part of his great chronicle of the times, Matthew Paris produced a five-page strip map showing the pilgrim route from London to southern Italy. This particular section covers the route from London to Dover via "Rosa" (Rochester) and "Cantuar" (Canterbury). The strip map's purpose was purely illustrative—it was certainly not intended to be a practical guide for travelers to follow.

24 The Gough Map, England, c. 1360

An unknown cartographer compiled this map, the oldest surviving route map of Britain, in the mid-14th century. Such was its accuracy that it was still in use 200 years later. Routes between towns are marked in red with distances given in Roman numerals. Towns are shown in some detail, important centers such as Bristol, Chester, and Winchester being lavishly illustrated, while the lettering for London and York is colored gold.

Of the four maps in question, the one in the *Abbreviatio Chronicorum* is the most carefully composed (see p.50), although the more sketch-like map in the *Historia Anglorum* is more accurate. Why this should be is unclear, although some of the problems may be related to the narrower width of the former map. Lack of room, for instance, could account for many of the misplacings of names, such as the positioning of Colchester, St. Osyth, and the name of Essex south of the River Thames. It is as if Paris believed it better to put a name in the wrong place than to have no name on the map at all.

These were not the only maps included in the *Chronica Majora*. All three volumes contained a five-page strip map, showing the route from London to the south of Italy (see p.51). An earlier attempt at creating a similar map was made by the cleric Giraldus Cambrensis, who spent much of his adult life hoping to become Bishop of St. David's in Wales, only to be disappointed when the Archbishop of Canterbury vetoed the appointment. The map of Europe Giraldus created in around 1210 reflects this disappointment, for, although it identified eight bishoprics in the British Isles, both Canterbury and St. David's are notable for their absence. To accompany the map, Giraldus produced a regional gazetteer plus an account of his personal experiences as a traveler to Rome. He made the journey four times in all: three to plead his case to be appointed bishop and the fourth, later in his life, as a pilgrim.

The Gough map

Apart from the single route he mapped from north to south, no roads appear on Paris's maps of the British Isles. This type of mapping had to wait until 1360, when the first detailed route map of Britain made its appearance. Drawn in pen, ink, and colored washes on two skins of vellum, the map gets its name from Richard Gough, an 18th-century antiquarian who bought it at a sale

25

Pinchbeck Fen, England, c. 1430
Local maps were comparatively rare in medieval times, so this one—probably drawn to help to establish which villages had the rights to graze sheep on the fen—is an exceptional survival. The identity of the man who drew it remains a mystery, but he was almost certainly a monk. He probably came from Spalding Abbey, Lincolnshire, since this is the area depicted in the greatest detail. The fen itself was colored a vivid green, while the churches—28 in all—were drawn in careful perspective and outlined in red. No secular buildings were shown.

in 1774. Its actual origins are mysterious. The name of the mapmaker is unknown, but it is considered likely that it was compiled—or at least used—by clerks of the royal court of Edward III (1327–77) as an aid to effective administration, while it has also been suggested that its creation was a response to the needs of the merchants of the time, especially those concerned with the wool trade on which the national economy was coming to depend. Maybe it was a reflection of the determination of the Plantagenets to assert their claim to a "monarchy of the whole island." Certainly, the map locates the chain of castles that Edward I (1272–1307) had had constructed along the northwest coast to enforce his subjugation of Wales.

The map is orientated with east at the top, and important rivers, such as the Thames, Severn, and Humber, are predominant. Other physical features are identified by symbols: a tree marks the location of the New Forest in southern England, for instance. Unlike other maps of the time, there are few written legends, but place names are ubiquitous. Routes between towns and cities are marked in red, together with the distances between them. Cartographically, many aspects of the map are remarkably accurate, particularly the area between Hadrian's Wall and the Wash, and the southeast of England, where the map is covered with towns and routes. Scotland, by contrast, has few settlements and no routes marked.

Local maps

Most local maps—those that exist were made largely in England and Italy—date from around 1400 onward, which suggests that only from about that time did medieval people as a whole come to appreciate the help that accurate maps could provide in the settling of land disputes and the like. One such map comes from Pinchbeck Fen in Lincolnshire, England, and dates from

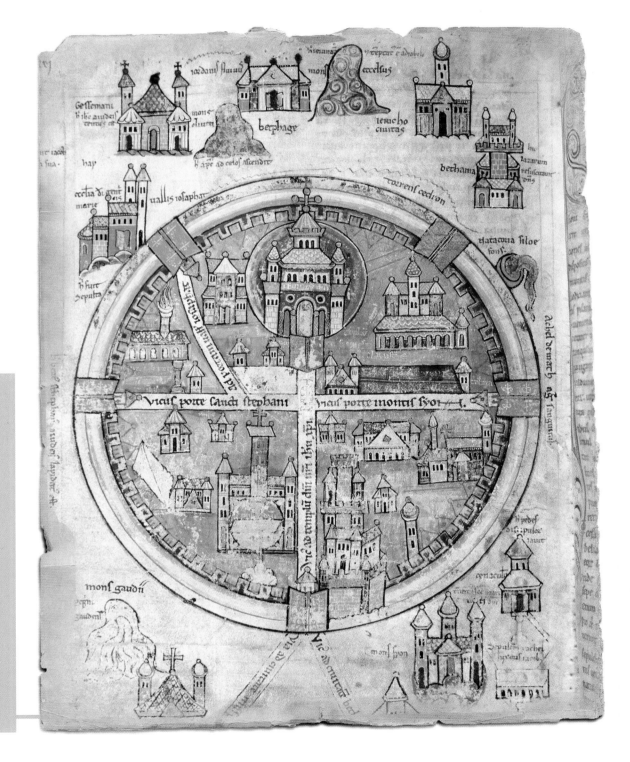

26 Jerusalem, England, c. 1120
Jerusalem was one of the few cities to be mapped extensively in Europe during the Middle Ages, even though for much of the time it was under Islamic control. This map was drawn to illustrate the *Historia Hierosolymitana*, a chronicle written by Robert the Monk to describe events leading up to and during the First Crusade. The city is shown in schematic form, with key buildings in some imagined detail. Maps such as this helped to keep Jerusalem, in which most Europeans would never set foot, at the center of the Christian consciousness.

around 1430 (see p.53). This map is thought to have been made at Spalding Priory, one of three important monastic foundations in the area, to record which of the six villages shown on the map had the rights to graze sheep on the common lands of the fen. Apparently five of the villages enjoyed such privileges, but one did not. The map showed only a small area of land—no more than 10 square miles (16 sq km)—but was packed with vividly rendered detail.

A somewhat similar map, thought to have been drawn in the mid- or late 15th century and showing Chertsey Abbey in Surrey, England, and the land surrounding it in sketch form, was intended to illustrate a similar argument, this time between the abbey authorities and the local farmers about who had the right to graze their cattle where. The map shows the abbey and its estates in detail, the nearby village of Laleham, a bridge across the River Thames, and two water mills that were owned by the abbey standing either side of the river.

CITY MAPS

Although only a few city maps survive from the medieval period, those that do are extremely interesting. Not surprisingly, Jerusalem features high on the list: at least 15 maps of the city were made in the 12th and 13th centuries, before it finally fell to the Muslims. It is known that a series of maps of northern Italian cities was drawn between 1291 and 1483, although medieval literature mentions the existence of city plans long before this. Charlemagne, the first Holy Roman Emperor (742/7–814), is recorded as having possessed his own maps of both Constantinople and Rome, but it is likely that, if these accounts are to be believed, such maps were survivals from the ancient world rather than medieval creations.

Despite this, it seems that there was relatively little interest in creating new, up-to-date European city and town plans until late in the Middle Ages. The situation lasted until the revival of interest in the accomplishments of the Classical world in the 15th century. In the 1440s, Leon Battista Alberti (1404–72) produced his *Descriptio Urbis Romae*, in which he outlined how best to draw a correctly scaled map of Rome through the use of angle measurements and paces.

Some time earlier, a detailed plan of Vienna had been compiled, although we have no idea by whom. What has survived is a mid-15th century copy of an original that was drawn in about 1422. The plan was based on the measurement of distance and was provided with a scale.

27 Constantinople, Cristoforo Buondelmonti, Italy, c. 1420

Buondelmonti, a Florentine nobleman, drew this plan of Constantinople as one of the maps to illustrate his *Liber Insularum Archipelagi* (The Book of Islands), an account of his exploration of the Aegean. His work was a fusion of geographical information, charts, and sailing directions: a prototype travel guide. The map shows the city's major landmarks clearly and was considered authoritative enough to be copied for the great atlas created for the ruler of Urbino over 50 years later.

Although there was no attempt made to set out the actual pattern of the city's streets, the map located the city's main buildings, indicating them in rudimentary and exaggerated perspective.

FRA MAURO'S WORLD MAP

It is fitting that this chapter should conclude with an account of one of the last world maps of the medieval period. It was the work of Fra Mauro, a Venetian monk from a monastery on the island of Murano, where he directed what was effectively a cartographic workshop. In response to a commission from the Venetian Senate and aided by Andrea Bianco, one of the most prominent portolan chart-makers of the day, Mauro set about his great task in about 1448. The map was completed by 1453. A copy was made for King Alfonso V of Portugal but it was lost, possibly on its way to the royal court in Lisbon. Mauro had already set to work on a further copy for the Venetian Senate, but he died before its completion. Bianco had to take over the work.

The map we have was undoubtedly one of the greatest maps of its day: not for nothing was Fra Mauro hailed as "geographus incomparabilis." It was immense, with a diameter measuring 6 feet 4 inches (1.9 m). Details were picked out in lapus lazuli and gold leaf. The map's unique combination of aesthetic magnificence and highly specific and useful information served to confirm Venice's rising status as a powerful political and commercial power.

Although Mauro preserved some of the characteristics of a traditional *mappa mundi*—the use of a circular format and the pictorial representation of cities, towns, and castles, for example—the rest was fresh and new, based on the latest geographical information. The map was orientated with south at the top, following Arab practice. Jerusalem no longer held center stage, while the customary decorative features and flourishes were banished, with the sole exception of a rendering of Paradise. There were no strange monsters to fill up blank spaces: Mauro firmly and realistically left what he did not know as *terra incognita* (unknown land).

For Europe, Mauro used portolan charts to help him to draw up improved coastal outlines of the Mediterranean and Black Sea areas, while his rendering of the far north of the continent was vastly superior to what had gone before. A legend on the map refers to the shipwreck of Pietro Querini, a Venetian explorer, on the Norwegian coast in 1431, and it has been suggested that Mauro obtained oral information on Scandinavia from him. For the west coast of Africa, Mauro had access to Portuguese charts—the Portuguese had charted as far as the Cape Verde Islands by 1445—while Arab sources supplied him with the information he needed to map the east coast. For information about the northeastern interior, he turned to emissaries from the Coptic Church, who, as he recorded, "with their own hands had drawn for me all these provinces, cities, rivers, and mountains with their names." To the east, the Indian Ocean was depicted as an open sea, while for the first time, China's two main rivers were charted with some degree of accuracy. The map was also one of the first in the West to represent Japan, a part of which—probably Kyushu—appeared beneath the island of Java.

As a whole, Fra Mauro's map was more up to date than the versions of Ptolemy's world maps that started to be circulated two decades later. Mauro was aware of and used his great predecessor's work, but he did not copy him slavishly. He stated that he was not being "derogatory to Ptolemy if I do not follow his *Cosmographia*, because, to have observed his meridians, parallels or degrees, it would be necessary to leave out many provinces not

mentioned by him." Fra Mauro's goal was simply to establish what he believed to be the truth. As he wrote:

> **"** *In my time I have striven to verify the writings through many years' investigation and intercourse with persons worthy of credence who have seen with their own eyes what is faithfully set above.* **"**

There could be no better epitaph to the medieval era than this precursor of the Age of Discovery that followed.

28

Mappa Mundi, Fra Mauro, Italy, 1459
Created by a Venetian monk known as Fra Mauro, this *mappa mundi* is far more advanced than any other maps of the time. As well as drawing on the rediscovered works of Ptolemy, portolan charts, and existing world maps, Fra Mauro took full advantage of explorers' accounts of their voyages and discoveries.

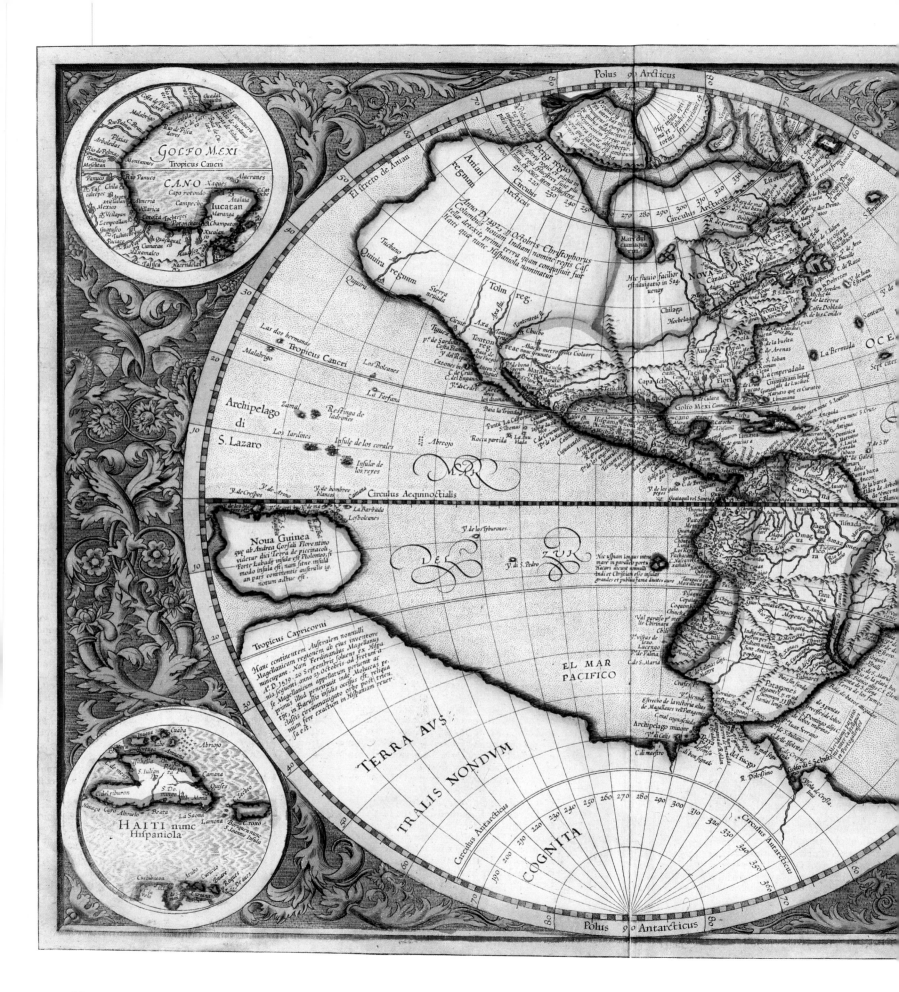

THE AGE OF DISCOVERY: Expanding Horizons

What has long been labeled as a great age of European discovery had momentous consequences—not simply for Europe, but also for the rest of the globe. It was the age when European sailors and ships ventured out from the coastal waters of the Old World to embark on what has been termed one of the greatest adventures in all human history.

Why this happened, when, and where is a matter of considerable historical debate. It coincided with a shift in the balance of economic and political power from the Mediterranean region to the Atlantic seaboard, with the gradual emergence, in turn, of Portugal, Spain, France, and England (to be joined by the Dutch, as their political independence was secured) as the dominant forces. The main reason for the voyages was often to establish trading relationships. One motivating factor might well have been religious: the desire to save souls, while another, which would have been in tune with general Renaissance thinking, was, to put it at its simplest, a growing natural curiosity.

Whatever the reasons, one of the consequences of the great explorations was that interest in maps and mapmaking greatly increased. Seafarers, who traveled further than ever before—especially across what one chronicler imaginatively christened the "green sea of darkness," the Atlantic Ocean—worked together with chart-makers and cartographers to expand European knowledge of the world. At the same time, the rediscovery of Ptolemy's *Geography*, with its translation into Latin from Greek in 1406, major innovations in navigational and surveying instruments, and the coming of printing, revolutionized the way in which many European maps were produced.

Their newfound knowledge of Ptolemy meant that Renaissance mapmakers now had access to the secrets and

29 The Americas, Michael Mercator, Germany, 1595

It took Gerardus Mercator (1512–94) over 20 years to create the maps for his great atlas—in fact, he never finished them. When his atlas was published in 1595, a year after his death, the atlas included maps by younger members of the family. His grandson Michael, for instance, drew this map of the Americas. Its compilation marked a turning point in European understanding of the geography of the New World.

skills of cartography based on mathematical principles, which included the use of map projections as a basis for reducing three dimensions to two and the employment of mathematics to calculate positions more accurately. Improved navigational instruments, such as the astrolabe, along with the appearance of books of navigation tables, enabled explorers to chart their voyages and discoveries more accurately than it had ever been possible to do before.

For its part, printing was to affect dramatically the way maps looked. It was also to transform the manners in which they were produced and distributed. Until the 15th century, many maps were precious objects, often available only to those with wealth and power. The coming of printing from moveable type, a result of the innovations of Johannes Gutenberg (c. 1400–68), began to change all that as far as Western Europe was concerned. Printing began to make maps less expensive and more accessible. Once a map had been transferred to a printing block or plate, it could be reproduced repeatedly, allowing the results to reach a new, wider audience— and to change their horizons.

THE PRINTING REVOLUTION

The first Western printed world map—a woodcut illustrating the geography section in the *Etymologiae*, a compendium of knowledge originally compiled by Isidore of Seville in the 7th century C.E.—appeared in Augsburg, in Germany, in 1472. It was rapidly followed by one created for the *Rudimentum Novitiorum* (A Handbook for Beginners), an illustrated guide to world history published in Lubeck in 1475. Both of these maps were traditional T-O maps.

Ptolemy's *Geography* was among the first cartographical reference works to be printed, in Vincenza in 1475, and it was a great success. The maps of the 1477 edition were based on Ptolemy's instructions and specially compiled by Pietro Bono and Girolamo Manfredi, two of the most celebrated Italian mapmakers of the day. Six editions had been printed by 1500, in Vincenza, Bologna, Rome, and Ulm. Not only was this a testament to the popularity of the original text and the new maps that were being created to accompany it, but was also an indication of how rapidly printing technology was spreading through southern and northern Europe.

Unlike the world maps in Ptolemy's *Geography* or the *Rudimentum Novitiorum*, which were addressed to scholars and students, Hans Rüst's printed map of the world, which was probably produced in Augsburg in around 1480, was meant for a wider audience. Not only was it relatively inexpensive to buy—there is a theory that it may even have been given away as a promotional item at fairs—but the annotations accompanying the map were printed in German, rather than in Latin. This probably made it the first map to be printed in the vernacular, thus reaching, and broadening the horizons of, a whole new audience.

The way to Rome

In 1500, Erhard Etzlaub (c. 1460–1532), a noted Nuremberg compass-maker, compiled his cartographic masterpiece: a map that he titled Rom Weg (The Way to Rome). According to contemporary accounts, Etzlaub was "admirably learned in the principles of geography and astronomy." The map was probably the first route map ever to be printed and, like the Rüst map, the text was in German rather than in Latin. Also like the map by Rüst, it was intended for popular use, primarily by pilgrims to Rome. The area the map covered extended from Scotland and Denmark in the north to central Italy in the south, and from Paris in the west to Poland and Hungary in the east. Rome was positioned close to the top of the map, with north at the bottom.

Printing workshop, French illumination, c. 1490
The coming of printing began to revolutionize the way in which maps were produced in the West. In southern Europe, maps were often engraved, while northern Europe preferred woodcuts.

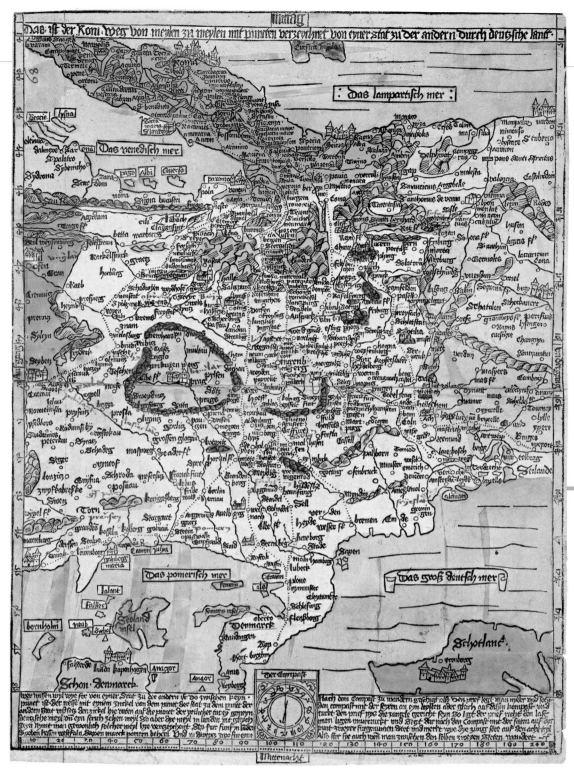

30

The Way to
Rome, Erhard
Etzlaub, Germany,
c. 1500

Nuremberg compass-maker Erhard Etzlaub compiled this map as a guide for the thousands of pilgrims flocking to Rome to celebrate the 1,500th anniversary of the birth of Christ. It was the first route map to be printed. Because it was intended for use by everyday travelers rather than scholars, the text of the map was printed in German, rather than in Latin.

On the map were depicted eight main routes—each of which was indicated by a dotted line—all converging on the eternal city. The dots were a form of scale, since each of them was intended to be the equivalent of a German mile. Over 800 towns were marked, signified by open circles, or, in the case of pilgrimage sites, by miniature pictures of churches. Etzlaub devised a new projection that allowed compass orientation according to stereographic projection, a precursor of the famous Mercator projection (see p.81). The map's preparation is thought to have been sparked by the coming celebrations to mark the 1,500-year anniversary of the birth of Christ that were to be held in Rome in 1500.

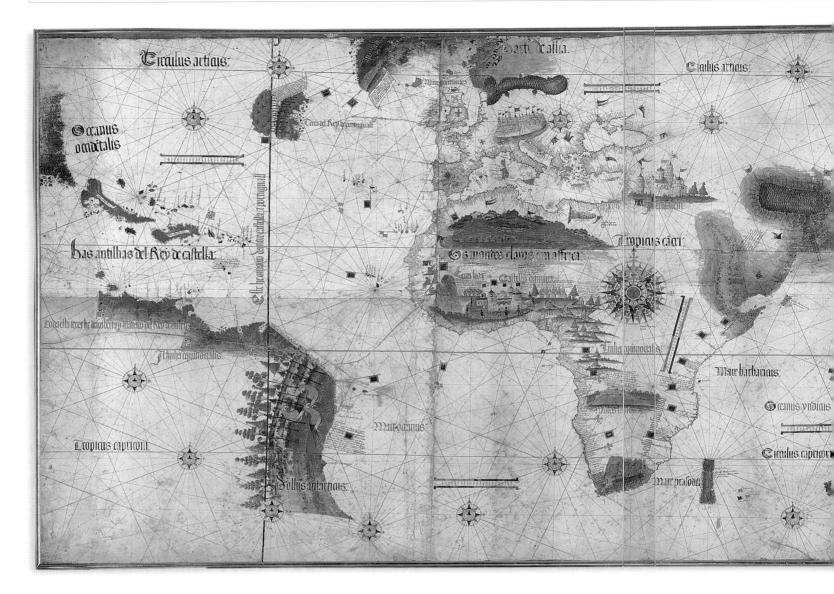

PORTUGUESE DISCOVERIES

From the time of the earliest maps until the present day, the ability to map land has been associated with the ability to control it. With the growing complexity of European politics, many rulers took an increasing interest in cartography: a well-mapped territory could be a well-governed and powerful one. Maximilian of Austria, who ruled the Holy Roman Empire from 1493 to 1519, was one such ruler, numbering some of the leading cartographers of his day among his close friends. According to the Bishop of Chiemsee, the emperor could himself sketch an impromptu map of any part of the lands he ruled.

The European fascination with colonial exploration could be argued to have started with Prince Henry "the Navigator" of Portugal (1394–1460). In 1418 he set up a court and observatory at Sagres in the Algarve which attracted Jewish and Arab scholars, mapmakers, astronomers, navigational instrument makers, and ship-builders, and he inspired them to discover and map new lands. There is some debate about how formal this "school" actually was. However, Pedro Nunez (1492–1577), a Portuguese geographer, wrote that Henry's navigators were "well instructed and well supplied with the instruments and rules of astronomy and geometry which all mapmakers should know."

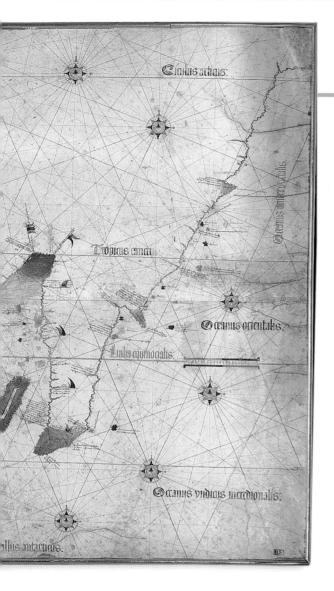

31

Cantino Planisphere, Portugal, 1502
Portugal wanted to keep its discoveries secret, but in 1502 Alberto Cantino, an Italian diplomat serving the Duke of Ferrara, obtained a copy of the master map prepared by the Portuguese hydrographical office. It is one of the earliest maps to show Columbus's discoveries. It is also the first known map to record the line agreed in the 1494 Treaty of Tordesillas, dividing land deemed to belong to Portugal from that belonging to Spain.

Portuguese exploration and mapping progressed hand in hand. From around 1415 onward, Henry organized and financed numerous voyages of discovery, with the aim of exploring an African trade route to India and the fabled spice lands of the Far East. After many failures, a Portuguese expedition rounded Cape Bojador, off the Western Sahara, in 1434, so opening up the possibility of trade with the Guinea coast. By the 1450s, Portuguese ships had reached the Cape Verde islands and the Azores had been settled.

Henry brought the cartographer Jácome de Majorca to Portugal, probably in the second or third decade of the 15th century. He was a precursor of the collection of professional geographer-navigators that the Portuguese crown employed in the second half of the 15th century. Their job was to produce maps quickly to incorporate new discoveries. The earliest of these maps known to survive is an anonymous chart dating from around 1471–82, and was followed by a chart by the Portuguese cartographer Pedro Reinel between 1484 and 1492. It depicts the extent of Portuguese maritime activity by this time, along the West African coast and into the Atlantic Ocean. Along the whole coast of West Africa, latitudes are correctly drawn, demonstrating the new usage of astronomical navigation.

After Henry's death, Portuguese seafarers continued the quest to reach Asia. In 1488, Bartholomew Diaz (1450–1500) reached what he christened the Cape of Storms, later to be renamed the Cape of Good Hope. It was this landmark event that conclusively put an end to the Ptolemaic notion that the Indian Ocean was a landlocked sea. It was Vasco da Gama (c.1460–1524) who finally succeeded in reaching India, when he made landfall at Calicut in May 1498. When de Gama returned triumphantly to Lisbon in 1499, his ships' holds were stuffed with spices, rare woods, and precious stones. It was enough to cover the costs of the voyage six times over.

The next year, Pedro Alvares Cabral (1467–1520) became the first European to sight the coast of Brazil, which he claimed for Portugal (naming it Ilha de Vera Cruz or "Island of the True

Cross") before retracing his steps eastward. The discovery was accidental: Cabral had been trying to follow da Gama's route to India, but storms and navigational miscalculations took him wildly off course.

The Cantino planisphere

The Portuguese crown considered the mapping of their discoveries sufficiently important to set up a government department to control the publication of their charts. This was the Armazém da Guiné e Índias (Guinea and India Office), which was responsible for the compilation of what the Portuguese called the *carta padrao del el-rei*. This was the map on which all new discoveries were officially recorded as quickly as possible after they were made, but neither it nor other official charts of the time seem to have survived. This may be because John II (1455–95) placed a ban on the distribution of such charts in an attempt to keep the new discoveries a state secret, or because they were lost in the great Lisbon earthquake of 1755.

One of the most famous cases of cartographic espionage ensued. In 1502, Alberto Cantino, a diplomatic agent employed by Ercole d'Este, Duke of Ferrara, paid an unknown Portuguese cartographer, or perhaps an Italian artist working in Lisbon, to copy the *padrao* secretly, which was then smuggled out of the Portuguese capital and back to Italy. Not surprisingly, the Portuguese were furious about the leak. They were determined to preserve their monopoly on the new geographical knowledge they were gaining or, at the very least, control its dissemination. The following year, Angelo Trevisan, the secretary to the Venetian ambassador to Spain, replied in response to a request for information about the Portuguese expedition to Calicut that it was "impossible to procure the map of that voyage because the king has placed a death penalty on anyone who gives it out."

The resulting map is known as the Cantino planisphere (see pp.62–3), and it is a fascinating record of Portuguese geographical knowledge of the world in the year 1502. To the west, the map shows what would later be rechristened Brazil in greatly enlarged form, its coast decorated with Portuguese flags, beautiful parrots, and lush trees. The map provided the Italians with knowledge of Brazil's coastline and of much of the Atlantic coast of South America long before other nations even knew that South America extended so far to the south. The African coasts and those of India are depicted with remarkable accuracy, bearing in mind that da Gama had rounded the Cape of Good Hope less than five years previously. India is drawn as a sharply tapering triangle with numerous towns marked on its western coast together with legends detailing the sources of their wealth.

The Cantino planisphere was also the first known map to record the line agreed in the 1494 Treaty of Tordesillas. After Columbus's discoveries in 1492, the Portuguese had felt threatened and complained to Pope Alexander VI that the Spanish had broken the terms of a treaty previously agreed between the two powers. The result was the Treaty of Tordesillas, in which all territory west of a line "drawn straight from pole to pole at a distance of three hundred and seventy leagues west of the Cape Verde islands" was deemed to belong to Spain, while all territory to the east was allocated to Portugal.

The Behaim globe

Another important record of the extent of Portuguese cartographic knowledge at the end of the 15th century is not a map or chart at all. It is a terrestrial globe: the oldest known European globe. Interest in terrestrial and celestial globes and the armillary spheres associated with them (devices used to demonstrate the movement of celestial bodies around the Earth or the Sun) increased in Europe from the late 15th century onward.

The Behaim globe was created by Martin Behaim (1459–1507), a cloth merchant from Nuremberg who arrived in Lisbon in the early 1480s and lived there for several years. Whether Behaim actually sailed on Portuguese expeditions to West Africa, as he claimed, is not certain, though we do know that, shortly after his arrival in Lisbon, he was invited to join the Junta dos Mathematicos, a group of Portuguese pilots and cosmographers who had been given the task of finding a way of using solar observation to calculate latitude more accurately. On his return to Nuremberg in 1490, he impressed his fellow burghers enough for them to commission him to create his great globe. It involved him in more than two years' work, with the help of painters and other craftsmen. It was finally completed before his return to Lisbon in 1493, probably in the preceding year.

The globe was a masterpiece. Particularly notable was the way in which Africa was depicted. His globe did not show the continent in the traditional way, as a continuous landmass running into Asia. Instead, the Erdapfel (Earth Apple), as Behaim termed his creation, drew on his knowledge of Portuguese exploration—especially of Bartholomew Diaz's voyage around the Cape of Good Hope in 1488—to show that it was possible to sail around the southern tip of the continent into an open Indian Ocean.

The globe contained a vast amount of information: 1,100 place names, 48 flags (10 of which were

32 Behaim Globe, Martin Behaim, Germany, c. 1492

Nuremberg merchant turned geographer Martin Behaim completed his globe—the oldest European globe to survive—in about 1492. Behaim derived much of his information from the writings of Ptolemy and Marco Polo; he also seems to have had access to first-hand Portuguese accounts of their voyages down the west coast of Africa and around the continent's southern tip. It is possible, though this is disputed, that Behaim himself went on such a voyage while resident in Lisbon.

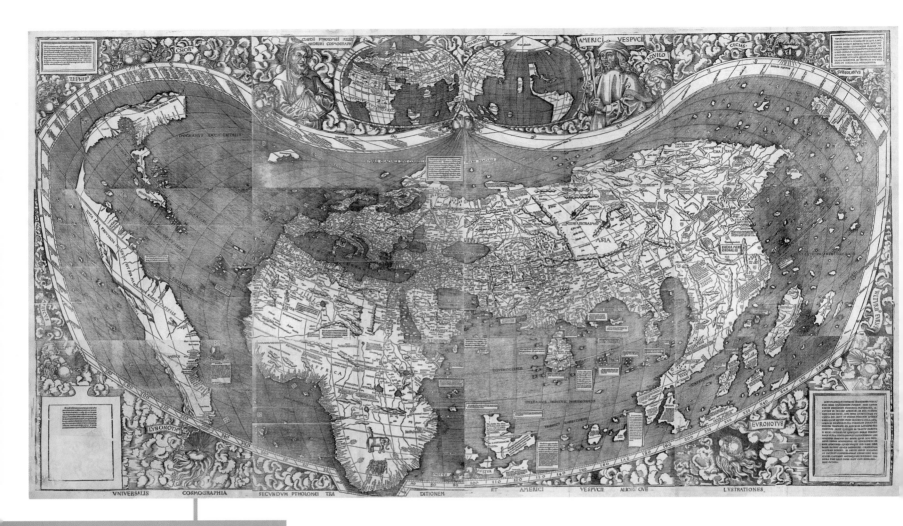

33 World Map, Martin Waldseemüller, Germany, 1507

Waldseemüller's creation measured 4 feet 6 inches x 8 feet (134 x 244 cm) and was the very first map to name "America." The name honored the Italian explorer Amerigo Vespucci, the first European to realize that the New World was a new continent. Although it was one of the most celebrated maps of its day, only a single copy is known to have survived. In 2003, the Library of Congress bought it for $10 million.

Portuguese), 15 coats-of-arms, and 48 illustrations of rulers. Eleven ships can be seen floating on the seas, which Behaim decorated with fish, seals, seahorses, serpents, and merpeople. Behaim also included commercial annotations, such as a list of where the best spices were to be found and a brief guide to the working of the spice trade. The annotations were intended to provide information both for traders planning voyages to the east and for investors when considering whether to give such voyages backing. The globe was both a remarkable product of, and a spur to, the dawning age of intercontinental trade. Its craftsmanship was not to be improved upon until nearly two hundred years later, when Vincenzo Coronelli created his terrestrial globe in Venice in 1688.

EUROPEAN DISCOVERY OF THE AMERICAS

At the same time as Behaim's globe was being completed, news of Christopher Columbus's momentous discovery of land in the western Atlantic was beginning to spread through the royal courts of Europe. Columbus (1451–1506) was a Genoese navigator who, apparently as early as 1481, had conceived the idea of reaching India by sailing west rather than east. For a long time, he was unsuccessful in finding anyone to back him—his brother approached the Portuguese, the French, and the English on Christopher's behalf, but all rebuffed him—until he offered his services to Ferdinand and Isabella, joint rulers of the Spanish kingdoms of Aragon and Castile, which formed the basis for a newly united Spain. After further delay and much debate, they agreed to Columbus's plan, granting him the impressive-sounding title of "Admiral of the Ocean Sea" and a host of trading concessions in any new lands he might discover. It is not known why the Spanish monarchs decided to back Columbus, but the most likely explanation is that they had decided to challenge the Portuguese in their attempt to monopolize trade with the spice markets of the orient, before the latter succeeded in establishing a sea route to the east.

Columbus made four voyages across the Atlantic in all: in 1492, 1493, 1498, and 1502. The first two took him to the Bahamas, where he made his first landfall, and then to Cuba and the island he christened Hispaniola (modern-day Haiti and the Dominican Republic), as well as to other parts of the Caribbean. The third also went to the Caribbean but ended in personal disaster. On his arrival in the Caribbean, Columbus discovered that the settlers of the colony he had founded in Hispaniola were rebelling against what they saw as his abuses of authority. They had him arrested and sent back to Spain in chains. Columbus's final voyage took him along the coast of Central America as far south as Panama. But to his dying day, Columbus believed that he had succeeded in his original intention. What he had discovered, he asserted, were islands off the coast of Asia, and sooner or later a passage would be found through or round them leading to China and India.

Another Italian explorer, the Florentine Amerigo Vespucci (1454–1512) served both the Spanish and Portuguese, and was Pilot Major of Spain from 1508 until his death from malaria four years later. On his Atlantic voyages, probably made in 1499 and 1501, he explored the northern coast of South America to well south of the mouth of the River Amazon and then sailed south to within a few hundred miles of Tierra del Fuego, South America's southern tip. Unlike Columbus, he concluded that what he had discovered was an entirely new continent. In a pamphlet titled *Mundus Novus* (The New World) he set out this claim, stating that:

> " *These new regions which I have searched for and discovered can be called a New World, since our ancestors had no knowledge of them... I have discovered a continent in those southern regions that is inhabited by more numerous peoples and animals than in our Europe, or Asia or Africa.* "

It is clear from this that, though Columbus might have been the first European to reach the New World, it was Vespucci who was the first to recognize it for what it was.

Amerigo Vespucci's New World

European mapmakers were quick to incorporate these new discoveries into their maps. Columbus himself might have drawn the earliest of these maps, which shows part of the northern coast of what is now Haiti. Vespucci, in contrast, drew no maps of his own, which is one of the reasons why his claims were heavily criticized by some of his contemporaries.

The first attempt to chart the new discoveries in detail was the work of Juan de La Cosa, a Spanish sea captain who is thought to have sailed with Columbus on his voyages across the

Illustration from a German edition of *Mundus Novus*, 1505 In 1503, Vespucci wrote an account of his voyage to the New World, in letter form, addressed to Lorenzo di Pier Francesco dei Medici. The work, known as *Mundus Novus*, was a great success: by 1510, there were as many as 24 editions.

Atlantic in 1492 and 1493. Dating from around 1500, de La Cosa's world map was conceived on a grand scale. In the far west, the map shows two vast jaws of land, between which the Caribbean islands are depicted in detail, stretching in a vast arc from Trinidad to Cuba. It is likely that de La Cosa used knowledge gleaned from Lucayan Indians to help him plot the Caribbean with such accuracy.

Like many other cartographers of the time, de La Cosa was not sure whether or not the transatlantic voyages had established the existence of a new continent. Martin Waldseemüller (c. 1470–1522), a cartographer based in Strasbourg, had read Vespucci's account of his discoveries and, when he came to prepare his world map in 1507, took what the Florentine explorer had written at face value. In *Cosmographiae Introductio*, a work that accompanied the map, Waldseemüller's collaborator Matthias Ringmann (1482–1511) wrote, "I do not see why anyone would rightly forbid calling it Amerige, that is, land of Americas, or America." This was how the newly discovered landmass was labeled on the map. It was the first map to depict the Americas as a distinctly separate continent, thus ending the European understanding of a world divided into only three parts—Europe, Asia, and Africa.

Waldseemüller's 1507 map is remarkable in several other ways, not least for its size (see p.66). It consisted of 12 separate sheets of paper, which measured 4 ft 6 in x 8 ft (134 x 244 cm) when laid out together. The map was the first printed map to cover 360° of longitude and the first to depict the Pacific as a distinctly separate ocean. As the map was published a decade before the first European was to sail on the Pacific, it is not known how Waldseemüller acquired his knowledge. He published his magnum opus six years prior to the historic first sighting of the Pacific by the Spanish conquistador Vasco Numez de Balboa in the course of his journey across the isthmus of Panama in 1413, and 15 years before the 18 survivors of Ferdinand Magellan's 1419 expedition returned to Spain, having accomplished the first circumnavigation of the globe.

Waldseemüller represented the Americas as two island-continents, both with a mountainous west coast. Beyond this some thousands of miles of ocean stretched west to the islands of Cipangu (Japan) and on to the Chinese coast. This was purely speculative. It is thought by some cartographic historians that Waldseemüller might have had access to sketches and papers drawn up by Amerigo Vespucci, but this supposition has not been verified.

We know only a very little about Waldseemüller's life and work, though we do know that, at the time of compiling the celebrated map, he was the official geographer to the court of René II, Duke of Lorraine. After 1516, he simply vanishes from the historical record. So, too, did his original map. It was rediscovered only in 1901, when Joseph Fisher, a Jesuit historian, found a copy of it securely bound up in an old book bearing the bookplate of Johannes Schirmer, a 16th-century German mathematician and geographer.

Christopher Columbus's New World

The only known view of the New World derived in part from a map prepared by Columbus (or so its creator claimed) is a map created by the Ottoman sailor and geographer Piri Reis in 1513. A native of Gallipoli, Reis is said to have begun his nautical career as a Barbary pirate, eventually rising to the rank of admiral (*reis*) in the Ottoman navy. He was beheaded at the age of 84 by order of his one-time patron Sultan Suleyman the Magnificent, following his failure to capture Hormuz from the Portuguese in 1554.

The map, which was lost until the late 1920s, is a fragment of a world map that has not survived in its entirety. Piri Reis wrote that he used Indian maps, Portuguese charts, and Christian

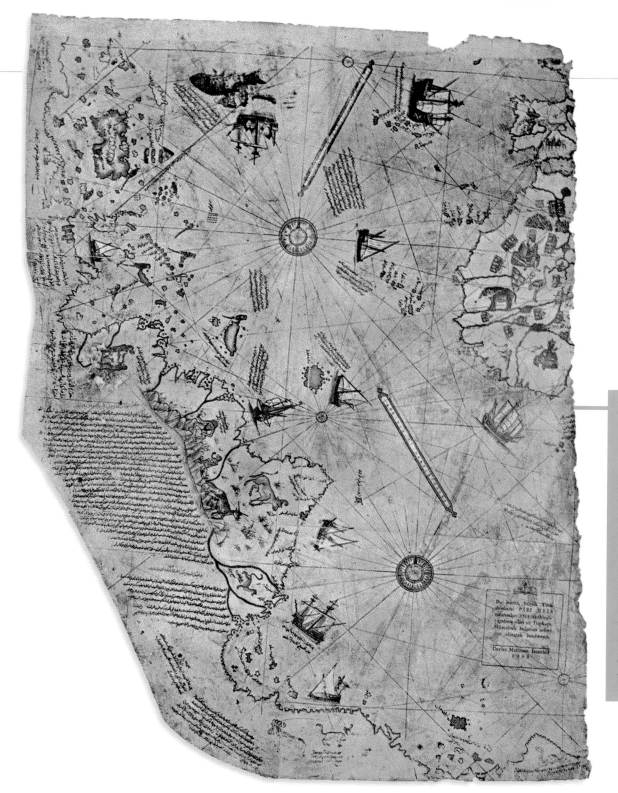

34
World Map, Piri Reis, Egypt, 1513
Piri Reis, one of the commanders of the Ottoman fleet and the man responsible for some of the finest maps and charts of the Mediterranean of the period, presented this world map to Sultan Selim I in 1517. Only a fragment of the western part of the map survives, but the notes Piri Reis appended to it demonstrate that he had access to a wide range of information. This included, so he claimed, "the map of the western lands drawn by Columbus."

mappae mundi as well as Ptolemy's maps to help him to compile it. From the surviving fragments, it appears that it was one of the most detailed maps of its kind to be produced in the early decades of the 16th century.

One theory about the fragment showing the New World is that it is a copy of a map taken from a captured Spanish sailor, who claimed to have sailed with Columbus on three of his voyages of discovery. Another suggestion is that it was derived from a map—long since lost—that Columbus made after his second voyage to the New World in 1493. The latter view is supported by a note that Piri Reis appended to his map, in which he stated that his work was "the result of comparison with…the map of the western lands drawn by Columbus, such that this map of the seven seas is as accurate and reliable as the latter map of this region."

CENTRAL AND SOUTH AMERICA

In 1508, the Spanish government, eager to regulate trade in the New World, set up the Casa de Contratación (House of Trade) in Seville. Following the earlier Portuguese example, its chief task was to keep a general map—the Padron Real—up to date and to ensure that Spanish sailors leaving on voyages of discovery were equipped with the best charts available. Soon the organization had at its disposal the finest cartographic talent in the Iberian Peninsula. The initial emphasis was on oceanic and coastal charts, but as the conquistadors journeyed deeper into the new continent, Spanish mapping of the New World evolved.

In 1518, the conquistador Hernando Cortés (1485–1547) set out to explore Mexico. After a series of battles, he established Spanish dominion over Central America and parts of South America: by 1522, the colony was larger than Spain itself. The Spanish colonization of the New

35

Map of Tenochtitlán, Mexico/ Germany, 1524

No one knows who drew this map of the Aztec capital, the drawings for which the conquistador Hernando Cortés appended to one of his letters to the Emperor Charles V in 1524. The style of the map, with its combination of pictures and plan, suggests that the drawings supplied to the Nuremberg engraver who prepared the printer's woodblock were either the work of Aztecs, or based on Aztec originals. The Latin text was added for the benefit of the emperor.

World was ruthless and relentless, and the colonies' most valuable exports, gold and silver, flooded back to Spain in ever increasing quantities.

The Spanish were largely destructive and dismissive of the local cultures and civilizations with which they met. However, in their journeys through the unknown terrain, the conquistadors found local mapping highly useful. Cortés himself almost certainly made use of the maps presented to him by the doomed Aztec emperor Motecuzama in his campaign of conquest as he marched southward from Mexico toward Guatemala. He is believed to have used other Amerindian maps along the way. Cortés's secretary, López de Gómara, praised Aztec maps as being reliable even beyond the boundaries of their empire. Later, the Spanish viceroy Antonio de Mendoza actively encouraged the production of indigenous maps—he termed these *pinturas* (pictures)—to help him in the task of settling land disputes and deciding on land grants.

The 1524 map of the Aztec capital of Tenochtitlán, as it stood before the Spanish destroyed it in 1521, is a classic example of how the two cultures, Spanish and indigenous, were drawn together when it came to mapmaking. In the map, Renaissance European and indigenous forms of mapmaking were employed, the result being a complicated juxtaposition of plan, bird's eye view, and pictorial illustration.

No one knows who drew the map. It almost certainly relied heavily on an indigenous source. We do know, however, that it was Cortés who presented the map to the Emperor Charles V in 1524. The map shows the main square—the Sacred Precinct, where the royal palace, the great temple, and the celebrated ball court stood—the causeways connecting the city and the mainland, and the aqueduct system, built by the Aztecs to ensure that their capital was supplied with fresh water. The place where Cortés set up his headquarters is marked with an oversized Spanish royal flag.

Indigenous mapping

The Aztecs were a mapmaking society. In common with other Amerindian cultures, such as the Mixtecs, Olmecs, Toltecs, and Zapotecs, they produced several different types of map, for different purposes. Díaz del Castillo, one of Cortés's lieutenants, recorded, for instance, that a "henequen cloth…on which all the pueblos we should pass on the way were marked" guided the Spanish forward. What del Castillo was referring to was, in fact, an itinerary map, and it is clear that the Aztecs produced many of these. The Aztecs' terrestrial maps made extensive use of hieroglyphs and pictograms, while their cosmographical maps divided the cosmos into three elements—earth, sky, and underworld—with the earth divided into four quadrants with the world tree in the center. Their celestial maps, which were far more accurate than the ones with which the conquering Spanish invaders were familiar, played a vital part in the creation of their sophisticated calendars.

Cartographic histories—maps of particular communities with accompanying historical narrative—were created throughout Mesoamerica. Even after the Spanish conquest, such histories continued to be drawn up by local indigenous communities. The Spanish christened them *lienzo* (canvas), because many of them were painted on sheets of cloth.

A typical Aztec example of such a work is known as the Codex Xolotl. Dating from 1542, after the Spanish conquest, it used maps of the Valley of Mexico as backdrops against which to tell the story of Xolotl, a legendary 13th-century warlord and his extended family. Topographic and hydrographic features are depicted. Another example, also thought to have been created in the early colonial period, tells the story of how the Chichimeques migrated from their original homeland in the north of Mexico to found Tenayuca. A path patterned with footprints indicates

Cortés entering Mexico, Mexican School painting, c. 1550
In 1519, the Spanish conquistador Hernando Cortés landed on the Yucatan Peninsula with about 500 men. By 1521, the great city of Tenochtitlán had fallen to the Spanish and their Amerindian allies.

36

This map was created after the
Spanish conquest by an
unknown Aztec cartographer to
form the first page of the so-
called Codex Xolotl, which tells
the story of the legendary 13th-
century warlord Xolotl and his
conquest of the Valley of
Mexico. As well as recording
the topography and
hydrography of the region,
the map gives numerous
hieroglyphic place names.

movement not just through space but also through time, from site to site. On entering the Valley of
Mexico, the emphasis becomes more planimetric. Towns, for instance, are now located on the
map as they were situated in relation to one another on the ground, while the various
topographical elements are also depicted with more geographical accuracy. The map reveals
an indigenous community recording its history in the face of European domination.

As far as is known, the pre-Hispanic societies of South America did not make use of paper or
of any other type of flat surface for recording purposes. This does not mean that the Incan culture
did not have a highly developed mapping tradition. The Incas, the rulers of an empire stretching
from modern Ecuador to Chile when the Spanish arrived on the scene, employed *quipu*,
elaborately knotted cords and strings, as a form of mapping.

Marking territory

By the late 16th century, the Portuguese, French, Spanish, and English were all claiming
land in the New World, and maps were made to show these claims. In 1562, a
cartographer of the Casa de Contratación, Diego Gutiérrez, drew up a map that charted the
Spanish and French landholdings in the Americas after the Peace of Cateau-Cambrésis had
been agreed between the two powers three years earlier. The map also set out the courses
of great rivers, identified many local tribes, and contained one of the earliest references
to California.

In 1571 Philip II of Spain gave the geographer and Cosmógrafo Real, Juan López de Velasco, the task of increasing Spain's knowledge of its overseas possessions. It was invaluable for the crown to understand the value and nature of the land it possessed thousands of miles away across the Atlantic. Velasco ordered the colonial governors to produce detailed geographical reports, *Relaciones Geográficas*, about the territories under their control, together with accompanying maps. A variety of draughtsmen created these *pinturas*, as they were termed. They were drawn up in a variety of styles: some showed only European influences, while others reflected indigenous cartography. Velasco also drew on the world atlas compiled by Alonso de Santa Cruz, his predecessor as Cosmógrafo Real, in 1542.

By the end of the 16th century, the Spanish had built up a detailed understanding of the main geographical features of their new possessions. In this they differed from the Portuguese, who, rather than colonize their new territories in Africa and the East, contented themselves with establishing a network of trading bases stretching from West Africa to China. These were little more than military outposts, set up to protect Portuguese commercial investments. It is notable that during the 1520s many of Portugal's cartographers defected to Spain, where opportunities to display their talents were more plentiful. Almost no Portuguese maps survive from the period of expansion that followed Vasco da Gama's great voyage, an epoch that saw the setting up of a

37 Chichimeque Migration, Mexico, 16th century

This map shows folk history rather than topography. It shows the Chichimeques (literally "wild people," so-called by the Aztecs) on the move to found Tenayuca in 1224. The map is typical of Mesoamerican migration maps of the time in that it relies on the journey it is recording—rather than the landscape—as a structure.

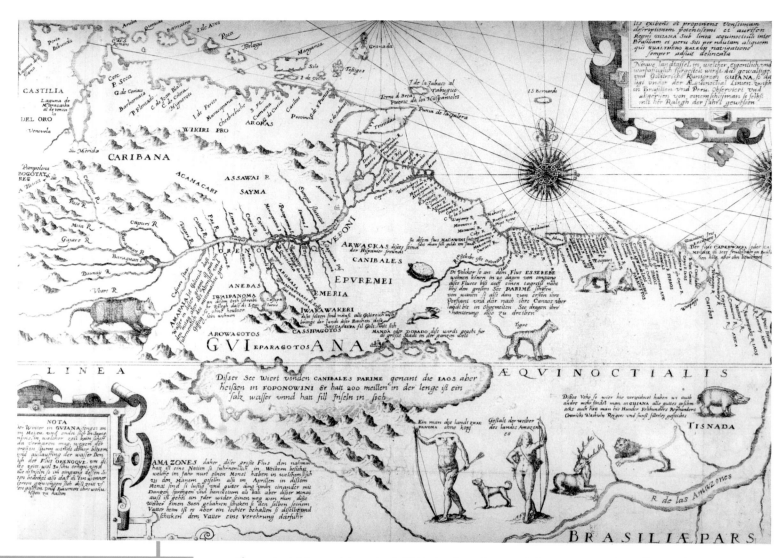

38 Guiana, Theodor de Bry, Germany, 1599

In 1590, de Bry undertook to document the European voyages of discovery with engravings, maps, and commentaries. This map illustrates the region of South America that Sir Walter Raleigh described as "Guinea." Most of the map was conjectural. Illustrations include a headless warrior.

settlement at Mozambique in 1505, the foundation of Goa in 1510 by Affonso de Albuquerque, and the establishing of a Portuguese presence at Malacca. Brazil was one of the few areas that the Portuguese settled extensively, and it was thoroughly mapped by João Teixeira Albernaz (c. 1575–1660), who produced six manuscript atlases of the country between 1616 and 1643, in which the number of maps increased from 16 to more than 30 as area after area was added.

Into South America

As in the Spanish conquest of Central America, the early European maps of South America were influenced to some extent by indigenous mapmaking techniques as well as relying heavily on geographical information obtained from local peoples, although such information was often misinterpreted. A classic example of such misinterpretation is on the map of Guiana prepared by the English explorer Sir Walter Raleigh (1554–1618) in around 1595. Raleigh was determined to locate El Dorado, the fabled lost city of gold, and the Amerindians he encountered on his explorations gave him the information he needed to draw a map accordingly. The mythical city is positioned firmly in the map's center on the shores of the equally mythical Lake of Parime.

A highly influential work on the Americas was the Flemish-born Frankfurt engraver Theodor de Bry's (1528–98) mammoth *Historia Americae sive Novi Orbis* (History of America or the New World), the initial nine volumes of which appeared between 1590 and 1602. De Bry based the maps and other illustrations on a variety of sources. Those for South America included the recollections of Hans Staden, a Germany mercenary in the service of Portugal who was captured by the Tupi Indians of Brazil; explorer Ulrich Schimdel's accounts of his travels to Brazil and Paraguay between 1535 and 1553; the missionary José de Acosta's *Historia Natural y Moral de las Indias*, which described Inca and Aztec customs in detail; and contemporary accounts of the two voyages that Raleigh was supposed to have made to Guiana (there is some doubt as to whether he actually went on the second one). The result was an early representation of the wonders of this part of the New World.

MAPPING NORTH AMERICA

European exploration and mapping of North America was largely the preserve of the English, French, and Dutch. The story started in 1497, when the Italian-born explorer John Cabot (1455–1498) sailed from Bristol under orders from Henry VII of England to sail to all parts "of the eastern, western and northern sea" to discover and investigate "whatever islands, countries, regions or provinces of heathens and infidels in whatsoever part of the world places, which before this time were unknown to all Christians." Within a month, he had reached Newfoundland, which he claimed for the English crown.

Having been acclaimed as the northern Columbus on his return to England (like his great contemporary, Cabot believed that he had found a new and shorter route to Asia), Cabot set sail again the following year, never to return. It is thought that he perished either from starvation after being shipwrecked on the Canadian coast or at the hands of local Native Americans.

In 1508, his son Sebastian Cabot (1481/2–1557) possibly explored the region to the north of his father's discoveries, seeking a northwest passage to Asia, and sailed down the eastern coast of North America. He transferred his allegiances to the Spanish, from 1512 to the late 1540s. It was while he was working in Spain that he drew up his celebrated 1544 world map, which includes detail of the Newfoundland coast. He returned to England in 1547, where he became a director of the Muscovy Company, set up to find a passage around the north of Russia to Asia.

Mapping New France

Where the English had led, the French followed. Between 1534 and 1541 Jacques Cartier (1491–1557) explored the region along the St. Lawrence River in what was to be christened New France and subsequently Canada. It is thought that he may well have been accompanied by what the French termed a *peintre* (literally, "painter") from the Dieppe school of cartographers. The maps of this school, which were produced for around two decades from the late 1530s onward, were notable for the information they gave about the latest discoveries and for their imagery, although according to Joao Laganto, a Portuguese pilot who was called into consultation by François I of France (1494–1547) while Cartier's third voyage was being planned, they were "well painted and illuminated, but not very accurate."

A member of the Dieppe school, Jean Rotz (born c. 1505) compiled his *Boke of Idography* in about 1540. Rotz had originally intended to present this work to François, but instead dedicated it to Henry VIII of England, who rewarded him by appointing him Royal

Hydrographer. He held the position from 1542 to 1547. Rotz's atlas—for that is what it was, although the word "atlas" had not yet been coined—covered practically the whole known world. It showed, in addition to outlines of the coasts and principal ports and harbors, the main activities of each area's inhabitants. In North America, for instance, he depicted somewhat unconvincing-looking Native Americans who appeared to be dressed like ancient Greeks, and embellished the map with what is likely to be the first European representation of a tepee.

Although the French royal family's interest in mapmaking in general continued after François' death, colonial mapping—and indeed exploration—languished until Henri IV (1553–1610) came to the throne. Henri was a noted map enthusiast: with his chief minister the Duc de Sully (1560–1641), he laid the foundations of state cartography in France, with the port of Le Havre at the mouth of the River Seine becoming the great center from which French overseas expansion was directed and recorded. Moreover, Henri was the patron of Samuel de Champlain (1567–1635), one of the greatest explorers and mapmakers of the day.

Champlain made his first voyage to the St. Lawrence River region in 1603, and he returned to North America on several subsequent occasions. His explorations took him throughout eastern Canada and New England, and he produced a number of maps along the way. These included a map of the east coast of North America from Cape Sable south to Cape Cod in 1609 and La Carte Géographique de la Nouvelle France, showing an area stretching from the mouth of the St. Lawrence as far upriver as the Great Lakes. Published in 1613, the map was intended to help with coastal navigation and to promote French settlement in the regions Champlain had explored, even though, in common with other explorers and mapmakers of the time, his knowledge was primarily of coastal areas, rather than of the interior. During the 17th century, French knowledge of the interior gradually increased as fur traders, explorers, and Jesuit priests in search of converts traveled up the St. Lawrence, west to the Great Lakes and south to the Mississippi River. The information secured by these travelers, much of it obtained from Native American sources, was used by Nicholas Sanson (1600–70), who taught geography to Louis XIII and Cardinal Richelieu, in his famous 1656 map Le Canada, ou Nouvelle France.

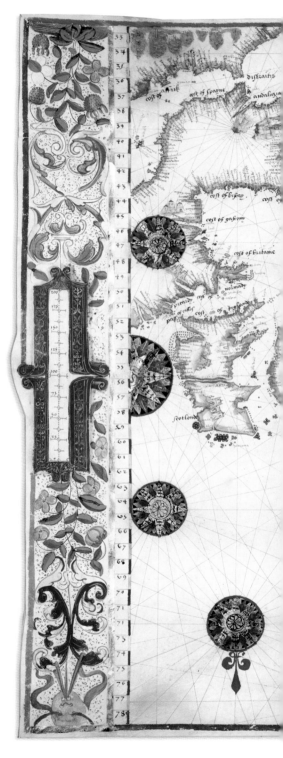

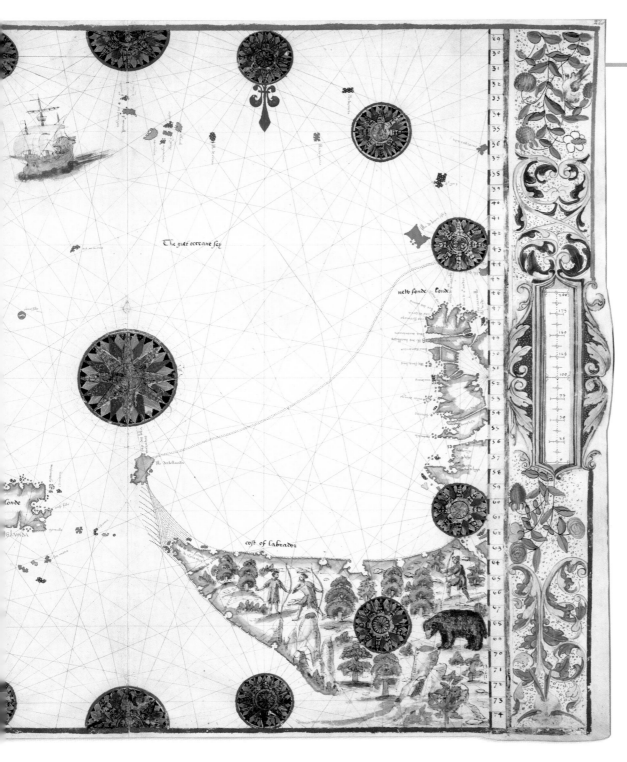

The great occeane sey

neh fonde londe

cost of labrador

Chart from the
*Boke of
Idrography*,
Jean Rotz,
England, 1542

French cartographer Jean Rotz
drew this map for the *Boke of
Idrography* he presented to Henry
VIII of England in 1542. The book
consisted of an introduction,
eleven regional nautical charts,
and a world map. This map shows
the North Atlantic Ocean, from the
east coasts of Labrador and
Newfoundland to Iceland, the
British Isles, and Spain. The
African, Asian, Indian, and Chinese
coasts are depicted with a
remarkable degree of accuracy in
the book, and even parts of what
was later known as Australia
appear. Rotz sailed on several
voyages of exploration. He might
have visited Sumatra in 1529 and
certainly went to Guiana and
Brazil in 1539.

English Virginia

During the reign of Elizabeth I (1553–1603) English interest in exploring and claiming territory in the
New World deepened. This was the era of the great Elizabethan seadogs—Drake, Grenville,
Hawkins, and Raleigh—who, surreptitiously encouraged by the queen herself, had begun to prey on
the Spanish treasure fleets transporting the wealth of Spain's overseas empire back to the motherland.

40 New France, Samuel de Champlain, France, 1613

Champlain's map of New France was the first to delineate in detail the New England and Canadian coast, from Cape Sable to Cape Cod. A number of habitations are shown along the shoreline: the larger ones represent French settlements and the smaller ones Native American villages. Anchors indicate where Champlain moored during his voyages.

Unlike the Portuguese, French, and Spanish, the English had no thriving home-grown school of cartography. In 1574, the mathematician William Bourne lamented in his *A Regiment for the Sea* that English sailors were forced to rely on charts made overseas, while two years later Drake himself had to visit Lisbon in order to buy the charts he needed for the voyage he intended to make around the world. He set sail in 1577, returning home three years later. Along the way, he landed on the northern coast of California, which he claimed for his royal mistress. For the next 200 years, the majority of mapmakers knew the region by the name Drake had given it: New Albion. As late as 1747, it was still widely thought to be an island rather than part of the American continent, an example of just how influential a cartographic error can be.

Given their lack of reliable charts, English privateers made good use of the maps of any ship they succeeded in capturing. Things began to change toward the end of the 16th century, when, in 1584, Sir Walter Raleigh persuaded Elizabeth to back his plan to found a colony in Virginia, a newly discovered territory that had been named after the Virgin Queen. The following year, Raleigh's cousin Sir Richard Grenville set up a short-lived settlement on Roanoke Island off the coast of Virginia. In 1587 a return expedition was led by John White, who turned out to be an accomplished cartographer, with Thomas Harriot as chief navigator.

On their arrival on the island, both men began to survey the coastline from Ocracoke to Cape Henry and westward up the Roanoke and Chowan Rivers. The map that White

produced as a result was a masterpiece, just as his paintings were a comprehensive survey of the flora and fauna of the region and of the life of the Native Americans he encountered. The map covered an area from Florida in the south to the Outer Banks, depicting the coastline and reefs in some detail.

John Smith and New England

Despite his ambitious ideas, Raleigh's colony was not a success. One of the great riddles of history is what became of the Roanoke settlers. When he returned with supplies in 1590, White found the settlement abandoned. Captain John Smith (c. 1580–1631), the man who gave New England its name, was fated to be similarly unsuccessful.

In 1615, having previously traveled to the New England and Maine area to survey the land, he set sail again, determined to establish the first permanent colony in the region. He never reached his intended destination. Just a few hundred miles out into the Atlantic, his ships were so damaged in a storm that all they could manage was to limp back to London. The indefatigable Smith tried again later the same year, this time being captured by French pirates. Although he eventually managed to escape from his captivity, the long-suffering investors on whom he relied for finance had had enough. They refused him more funds.

Smith's legacy was cartographic rather than territorial. Prior to his involvement with New England, he had spent some years in Virginia as de facto governor, where he led two surveying expeditions into Chesapeake Bay at the behest of the Virginia Company. He used what he had discovered as the basis for an extremely detailed map of the region, which was published in 1612. His equally ambitious map of New England "observed and described," which he produced in 1614 as part of his propaganda to secure the financial backing he needed for his next voyage, was based on sketches he had made sailing from Nova Scotia to Rhode Island in a small portable boat while his crew busied themselves fishing the Grand Banks for cod (see p.80).

As soon as it appeared, the New England map became extremely popular, and ran through many editions. On the map, Smith deliberately emphasized the region's Englishness, so making the land appear ripe for quick, easy settlement. Indigenous names, for instance, were suppressed, while there was no indication that Native American tribes inhabited the area. The verse accompanying the large portrait of Smith at the top left of the map boastfully sets out his

41 Virginia, John White, Virginia, c. 1587
In 1587, the artist John White led an ill-fated expedition to Roanoke Island in Virginia. Along with Thomas Harriot, the expedition's chief navigator, White surveyed the east coast of America in great detail. This map covers Florida in the south to the Outer Banks. It was the earliest known English attempt to map their possessions in the New World.

ambitious intentions as the presiding genius of the new colony. In 1620, the Pilgrim Fathers arrived in North America and established the first permanent European colony in New England.

REVOLUTIONARY MAPMAKERS

In the 16th century a cartographic revolution took place in the Low Countries, led by mapmakers such as Gemma Frisius (1508–58), Abraham Ortelius (1527–98), and Gerardus Mercator (1512–94). The reasons for this shift in the balance of cartographic power northward were largely economic and commercial. Antwerp, already one of northern Europe's most prosperous ports, had overtaken Lisbon as the European center of the spice trade, while Flemish and German bankers were underwriting many of the voyages of exploration and discovery then taking place at an ever-growing pace.

42 New England, John Smith and Simon Van der Passe, England, 1614

In the early years of the 17th century, Captain John Smith made several expeditions to North America. In 1614, he surveyed the Massachusetts Bay region and persuaded Charles, Prince of Wales, to christen the area New England. The map was probably drawn by Simon Van der Passe, the son of a Dutch engraver, who based his rendering on Smith's detailed sketches. The map was intended as a propaganda tool to promote English settlement of the region.

Frisius was head of the great school of geography and cartography at the University of Louvain, in the Low Countries, where the young Mercator enrolled as a pupil in 1530. Frisius wrote *Libellus de Locorum Describendorum Ratione* (Little Book on a Method of Delineating Places), in which he laid down the principles governing the use of triangulation in mapmaking. Ortelius, a native of Antwerp and a close friend of Mercator, was considered to be the foremost "painter of maps" of the day before he became a cartographer in his own right. As for Mercator, he is probably the best-known mapmaker of all time. He alone, it seems, had the imagination and intelligence required to solve a problem that had been baffling mapmakers for centuries. This was the devising of a method by which a rendition of the curved Earth could be achieved on a flat surface in a way that could be used for navigation.

Abraham Ortelius

From 1547, Ortelius was an illuminator of maps, and soon also traded in books, maps, and prints. He started making maps of his own in the early 1560s. In 1570 he published *Theatrum Orbis Terrarum*, which is considered to be the world's first atlas in the modern sense of the word: a collection of maps of the same dimensions and a survey of the world as known at that time. It started with a world map (see pp.82–3), followed by detailed maps of all the known continents: the Americas, Asia, Africa, and Europe. For the first and last of these, Ortelius probably turned to Mercator. The map of Europe closely followed the one produced by Mercator in 1554. As far as the New World was concerned, the map in the original edition, possibly derived from a lost map by Mercator, was replaced in 1587. The depiction of the western coast of South America was inaccurate, while an impressive—if nonexistent—continent made its appearance in the middle of the Pacific Ocean. North America, on the other hand, looked very much the same as it would on much later maps, with California being shown correctly as a peninsula.

The *Theatrum* was an immediate commercial success. Within a few years of its appearance, it had been translated from its original Latin into Dutch, German, French, Spanish, Italian, and English. There were 21 editions during Ortelius' lifetime, and 13 after his death. One of the reasons for this success, in addition to Ortelius' financial shrewdness in keeping every aspect of production and distribution firmly in his own hands, was the quality of the maps the *Theatrum* included. Ortelius was insistent that only the best available maps by the most celebrated and competent cartographers would be used for reference and that these debts would be properly acknowledged. Each edition was also updated. Whereas the 1570 original contained 53 map sheets, the Italian edition of 1612, for instance, was more than double the length.

Mercator's projection

It was Gerardus Mercator, Ortelius' Flemish contemporary, who first coined the term "atlas," to describe the great world survey that dominated his activities for the last decades of his life. He chose it, so he said, "to honor the Titan Atlas, King of Mauritania, a learned philosopher, mathematician and astronomer."

The publication of his *Atlas*, as it is quite simply known, was the climax of a long and distinguished mapmaking career, which began in 1537 when Mercator engraved his first map, a detailed view of the Holy Land. Printed on six sheets of paper, it made a sizeable wall map when glued together. He published his first world map the following year (see p.84), and other important maps swiftly followed, notably a 1554 map of Europe. Again engraved in copper but this time printed as 15 separate sheets, the map measured 47 x 58 inches (120 x 150 cm) in its entirety.

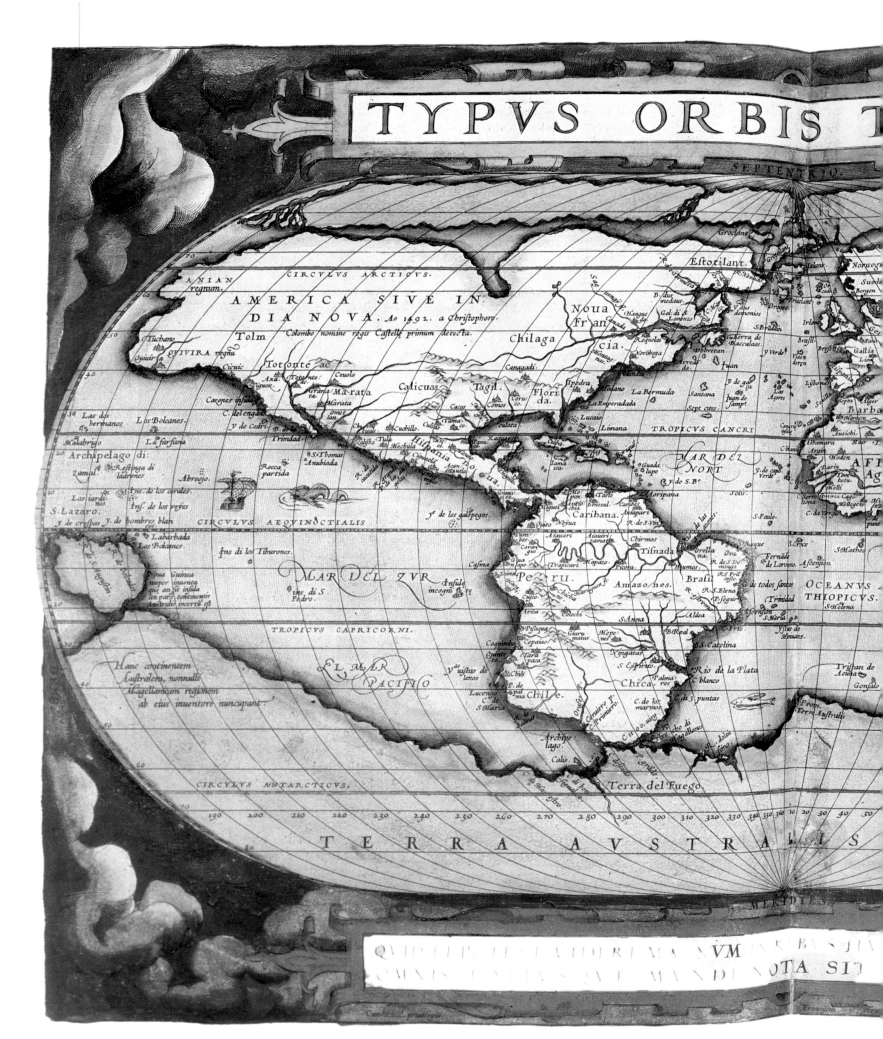

TYPVS ORBIS T

ANIAN regnum.

CIRCVLVS ARCTICVS.

AMERICA SIVE IN DIA NOVA. Ao 1492. a Christophori Colombo nomine regis Castelle primum detecta.

Groclant.

Estotilant.

Noua Franci cia.

Chilaga

Tolm

Tuchano

QVIVIRA regnu

Quuir a

Cicuic

Totonte ac

Canagadi.

MAR DEL NORT

TROPICVS CANCRI

Calicuas

Tagil.

Flori da.

Hispania

S.Thomas Anubiada

Rocca partida

Abreojo

Caribana.

CIRCVLVS AEQVINOCTIALIS

y de los galópegos

Ins. de los corales.

S.Lazaro

Labarbada
Los Bolcanes.

Ins di los Tiburones.

MAR DEL ZVR

Casma

Pe ru.

Amazones.

Brasil

OCEANVS THIOPICVS.

Ansule incogni te

ins di S. Pedro.

Noua Guinea nuper inuenta que an sit insula an pars continentis Australis incertu est.

TROPICVS CAPRICORNI.

EL MAR PACIFICO

Hane continentem Australem, nonnulli Magellanicam regionem ab eius inuentore nuncupant.

Chica

Rio de la Plata

Chile.

Archipe lago.

Calis.

CIRCVLVS ANTARCTICVS.

Terra del Fuego

190 200 210 220 230 240 250 260 270 280 290 300 310 320 330 340 350 360 10 20 30 40 50

TERRA AVSTRALIS

RRARVM.

Tazata

Taingin · Mongol.

Naiman · Mongul

Mongol.

Ocka · Wiliki · Calami · Coffin · Cattigara · Tenduc

Tartaria · Gruftina · Copi bálech · Kenai be · Turfon · Campi · Cathaio · Chainat · Gouza

Marmorea · Bulgar · Olg.A. · Cotam · Singui · Congu · Quanzu · Rohana

A S I A · Paianfa · Tafkent · Caratzan · Camdu · Tangu · Brema

Armenia · Mar de Bachu · Turcheltan · Samarchand · Voci · am · Iaci · Quinfai · Miaco

Natolia · Bachu · Coralan · Danim · China · Curfium

So Perfia · Thus · Rey · Baicu · del · Bonprun · Iatim · Corfi. · Ia. pan.

Ierufalé · Saura · Sirás · Turbut · Candahar · Serchis · Mien · Ala · Cantan · Xaion · Litampe · Lequio

Guzarate · India orien · talis · Lichi

Aegyptus · Qui libi · Calaiate · Diu · Delli · Tame · Deltam · Cai

Geogan · Hecha · Indu · Cambaia · Orixa · Brema · Tinhofa · Panodas · Humu

Nubia · Lacari · Zibit · Fartach · Aden · Goa · Narfiga · Cape Ian · Calicut · Pulo · Palohan · nu · han.

Abiffini · Aca · une · Magadaxo · Ravtora · Y di Maldiuan · Comari · Zeilan · Malaca · Pulo · furfur · Mocluce infule

con · Cagria · Melinde · Vasco de Acuña · Gifam · Mano · Iaua riá · upr. · Nunda · uno

Quiloa · S. Francesco · Due Compagne · Iaua riá · Campara · Banduar

Mozambique · Adarno · Liona · Don Garcia · Poueada · fona · Batuliar

Gebage · Baixos de Nazaret · Mascarenas · non galopes · Lantchidol · mare. · BEACH

Iulis Laurentius · S. Apollonia · Iuan de · Lisboa. · LVCACH · Iaua · minor

Burú · Punta de S. · Maria · Tomeri · MAR DI INDI · Petan

de las · nacas · Los Romeros · Saldo · MALETVR

Valtiffimas hic effe · regiones ex M. Pauli Ven: et · Lud. Vartomani fcriptis pe: · regrinationibus conftat ·

regió · llata ob in · vium Abidem ·

N DVM COGNITA.

VI AETERNITAS · TVDO. CICERO.

---

43 World Map, *Theatrum Orbis Terrarum*, Abraham Ortelius, Low Countries, 1584

In his masterly compendium, *Theatrum Orbis Terrarum* (first published in 1570), Abraham Ortelius brought together an unrivaled collection of maps of most of the known world. The work has often been described as the first modern atlas. Combining maps of uniform size and style with comprehensive text, the *Theatrum* set the standard for the shape and contents of future atlases. The work became the bestselling atlas of the 16th century, over-shadowing Gerardus Mercator's *Atlas* (1585), which was the first to coin the term. Ortelius' work ran into 34 editions and was translated into all the major European languages. Many of these editions included new or improved maps: for instance, a better map of America replaced the original one in 1587. The *Theatrum* spawned another first, the pocket atlas, entitled the *Spieghel der Werelt* (most editions were known as *Epitome Theatri Orteliani*), in which the maps were redrawn by Ortelius' collaborator, Philip Galle.

44

World Map, Gerardus Mercator, Low Countries, 1538

Just a year after publishing his first map—a cartographic portrait of Palestine—Gerard Mercator produced his first map of the world, in 1538. In compiling it, he employed the double cordiform projection pioneered by the French mathematician Oronce Fine in 1531. His map depicts the world as two heart-shaped views of a globe—one from the North Pole and the other from the South Pole. Mercator was careful to differentiate previously mapped coastlines from those in areas that were still largely unexplored. By 1569, when Mercator published his second world map, his methods had changed dramatically. He used his own newly devised, revolutionary map projection to compile it, still known today as the Mercator projection.

According to Walter Ghim, Mercator's first biographer and his neighbor in the German town of Duisberg, where he lived from 1552 until his death, the cartographer's map of Europe "attracted more praise from scholars everywhere than any similar geographical work which has ever been brought out." This was despite some obvious weaknesses, for example in the map's depiction of Britain. Mistakes in the rendering of place names (Norfolk, Suffolk, and Essex, for instance, are written as though they were villages, rather than counties) as well as some

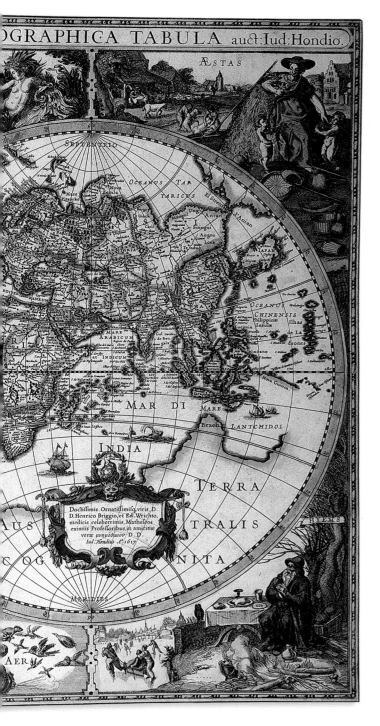

surprising omissions, such as Mount Snowdon and Windsor Castle, point to the mapmaker's lack of direct knowledge of the island. One of his main sources was almost certainly a map made by an English cartographer called George Lily and first published in Rome in 1546.

The map of England, Scotland, and Wales that Mercator created a decade later was far more accurate. The exact source of the new information, though, remained a mystery. All Ghim had to say on the subject was that "a distinguished friend sent Mercator from England a map of the British Isles, which he had compiled with immense industry and utmost accuracy, with a request that he should engrave it." No one knows for certain who this so-called friend was, but it is now thought that it was likely to have been a disreputable Scottish Catholic priest called John Elder. Apparently, Elder had access to the Tudor royal library, where he copied supposedly secret drawings made by English surveyors which he took with him when he was forced to flee the country in 1561. Perhaps Mercator's seemingly curious reluctance to acknowledge Elder's contribution—if, indeed, there was one—was a consequence of the religious turmoil of the times. As a Protestant, Mercator might well not have wanted it to appear as though he was in cahoots with a Catholic—and certainly not with one who turned out to be an active supporter of Mary, Queen of Scots and favored the overthrow of Elizabeth I.

It was in 1569 that Mercator produced a new type of world map that was to win him cartographic immortality. The innovation was not so much what the map showed, but how the information was displayed, for Mercator had devised a new and revolutionary map projection. He represented the world as a rectangle with the Arctic and Antarctic regions flattened out to the same degree as the equator. In his own words, he had "spread on a plane the surface of a sphere in such a way that the positions of all places shall correspond on all sides with each other both insofar as true direction and distance are concerned and as concerns true longitude and latitude."

Sailors found that the maps and charts that were soon being created according to Mercator's projection were invaluable. They made it possible to set a compass course more accurately than ever before. With the new projection, they could plot a line on a map that could then be reliably followed through the use of the appropriate compass bearings.

Mercator's atlas

It was while preparing his edition of Ptolemy's *Geography*, published in 1578, that Mercator began work on his world atlas. Advancing years, illness (he suffered two strokes in the early 1590s), the difficulty of resolving discrepancies between his sources, and the sheer amount of labor involved in engraving the plates delayed progress, so he published what turned out to be his last and greatest cartographic work in instalments. The 1585 edition contained 51 maps, mainly focused on the Low Countries, France, and Germany. The 1589 volume, with 23 more maps, extended the coverage to Italy and Greece. The complete 1595 edition reprinted the 74 maps that had been issued earlier with the addition of 33 new ones. These covered most of the remaining parts of Europe, leaving Portugal and Spain as the only areas lacking detailed coverage. Rumoldus, Mercator's third son, and three of his grandsons later added a world map, regional maps of Asia and Africa, and a map of America (see pp.58–9).

The *Atlas* was not a commercial success when it was first published, so Mercator's grandsons sold the plates to the Hondius family, who ran a successful publishing concern in Amsterdam. In 1606, they expanded the *Atlas* by adding another 37 maps, drawn and engraved by Jodocus Hondius (1563–1612). A more inexpensive edition with smaller maps was also published. This was christened the *Atlas Minor*, 25 editions of which appeared between 1607 and 1738. The original atlas, too, was reprinted successfully 30 times between 1606 and 1641, with the Latin commentary being translated into Dutch, French, German, and English. It set the standard for all the so-called "grand atlases" that Dutch cartographers produced during the remainder of the 17th century.

THE RISE OF DUTCH CARTOGRAPHY

By the 1600s, a new nation was fast becoming an important player on the global stage. It had taken decades of war, interspersed with periods of truce, for the Protestant citizens of what were now called the United Provinces to win their independence from Catholic Spain in 1609. The long conflict was reflected in the cartography of the period, as in the example of a propagandist Dutch world map, created in around 1596. The map depicts a *"Christiani militis"* (Christian knight) fighting against a host of evils, which the map's creator, Jodocus Hondius, symbolically identified with Catholicism and the Spanish foe. The face of the knight, it is thought, was drawn to resemble Henri de Navarre (later Henri IV of France), a Protestant hero of the time. Another map of roughly the same period depicts the United Provinces in the shape of a lion, a symbol of the House of Orange, fighting off the Catholic enemy.

By the early 17th century, the tentacles of Dutch trade were starting to wind their way around the Mediterranean and the Baltic, across the Atlantic, south to Africa, and across the Indian Ocean to India and on to Southeast Asia, where the Spice Islands lay. Realizing that their continued survival as a prosperous, powerful, and independent nation depended on the success of their sea-borne commerce and thus their knowledge and command of the sea, the Dutch

Title page, Mercator's ***Atlas sive Cosmographicae Meditationes de Fabrica Mundi et Fabricati Figura***, 1595
Gerardus Mercator was the first to coin the word "atlas" to describe a bound collection of maps. The title page of his 1595 edition shows the Titan Atlas cradling a globe.

succeeded in attracting the best mapmakers of the day to their new republic, many migrating north from the great cartographic center and port of Antwerp in modern-day Belgium, to Leiden, Amsterdam, and The Hague. It was in Leiden that Lucas Janszoon Waghenaer (c. 1534–1605) produced his *Spiegel der Zeevaert* (Mirror of the Sea), the first printed sea atlas, in 1584 (see p.88). By 1592, it had been published in five languages. The work covered the coastal waters of northern and western Europe in impressive detail and set the pattern for subsequent generations of similar sea atlases that would eventually cover the whole of the known world.

The Blaeus and their maps

For a time, Amsterdam became the world's mapmaking capital. It was particularly fitting that the Citizens' Hall of the city's 1655 Town Hall had an inlaid, double-hemispheric map of the world set into its marble floor as a testament to Amsterdam's commercial, maritime, and cartographic importance. The city was the headquarters of many leading mapmakers of the day, including Jodocus Hondius, Johannes Jansson (1588–1664), Nicholaus Jansz Visscher (1618–79), and Johannes van Keulen (1654–1717). The Blaeu family of mapmakers set up business in the city in 1599.

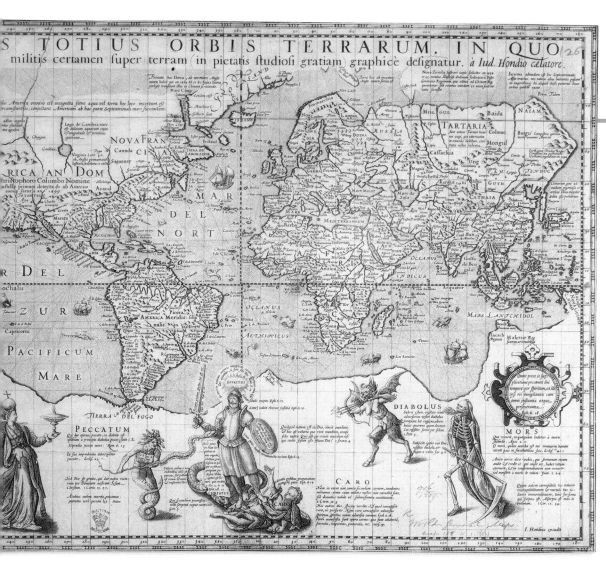

45

Christian Knight World Map, Jodocus Hondius, Netherlands, c. 1596

By the time Jodocus Hondius, one of the leading Dutch mapmakers of the day, created this world map, the Protestant Dutch had been fighting the Catholic Spanish for their independence for decades—and it can be seen from his map that Hondius strongly identified himself with the Dutch cause. The heavily armored Protestant knight, at the center bottom, is fighting bravely against "vanity," "sin," "carnal weakness," "the Devil," and "death," all of which are symbolically linked to Catholicism and the Spanish foe, making this a noteworthy example of cartographic propaganda.

46 Spiegel der Zeevaert, Lucas Janszoon Waghenaer, Netherlands, 1584

Waghenaer's great sea atlas—the first work of its kind to be printed—marked the beginning of a new era in chart-making. It coined the term "waggoner," which became synonymous with pilot guides. The atlas covered European coastal waters from Spain to Norway, with sailing directions and related information on the reverse of each chart. The chart shown here is of the southern coast of England.

The family patriarch was Willem Janszoon Blaeu (1571–1638), who served as the official hydrographer to the Dutch East India Company from 1633 until his death and was then succeeded by Joan Blaeu (1596–1673), one of his two sons. Willem Blaeu acquired the plates of Mercator's *Atlas* and used them to publish a world atlas in 1630. This was followed by the *Novus Atlas* or *Theatrum Orbis Terrarum* in 1635, published in two volumes and containing 208 maps (see pp.90–1). More maps were added over the years and, by 1655, it was six volumes in extent. From 1662, Joan Blaeu published the 3,000-page, 600-map *Atlas Major*, which

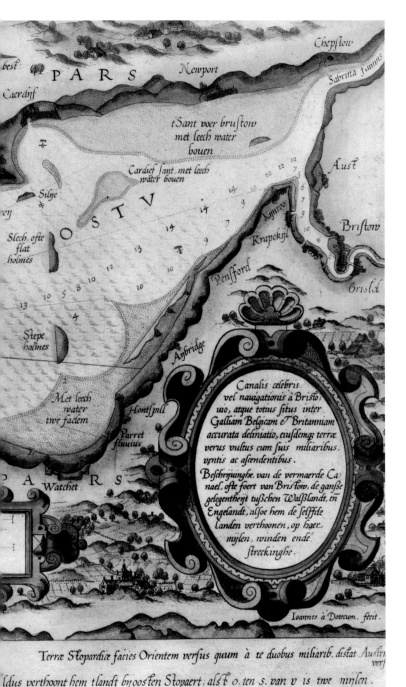

marked the high point of 17th-century Dutch cartography.

The New Netherlands

The Dutch were not only traders, although this was obviously their priority: they were also colonizers. In 1609, the Dutch East India Company sponsored the English explorer Henry Hudson (d. 1611) in an attempt to locate a northeastern passage to what he termed the "islands of spicery" in the East Indies. Having been forced to turn back in the face of inclement weather and a crew which shared his lack of enthusiasm—and not wishing to face his Dutch paymasters with nothing to report—Hudson ventured west to the Maine coast, where he sailed south and then north again in search of the equally elusive northwest passage.

Eventually, Hudson reached the mouth of a wide river, later named the Hudson River, and continued up it for 150 miles (240 km) until he reached the location of present-day Albany in New York State. There the water became too shallow for him to continue. His voyage back to the Netherlands was unlucky. Putting into Dartmouth on the Devon coast of England, he was arrested and charged with entering a foreign service without the permission of James I, the English king. His logbooks, together with the maps and charts he had made, were confiscated.

Although Hudson never returned to the Netherlands (he died on a voyage in 1611 when his crew mutinied and cast him adrift), reports of his discoveries reached Amsterdam. The Dutch East India Company sponsored Adriaen Block (1567–1627) to revisit and map the area. They were attracted to this new region because no other European power had established a presence there. It was also rich in beaver, an extremely valuable commodity. In 1614, Block, his fellow captain

ASIA
noviter delineata

Auctore
Guiljelmo Blaeuw.

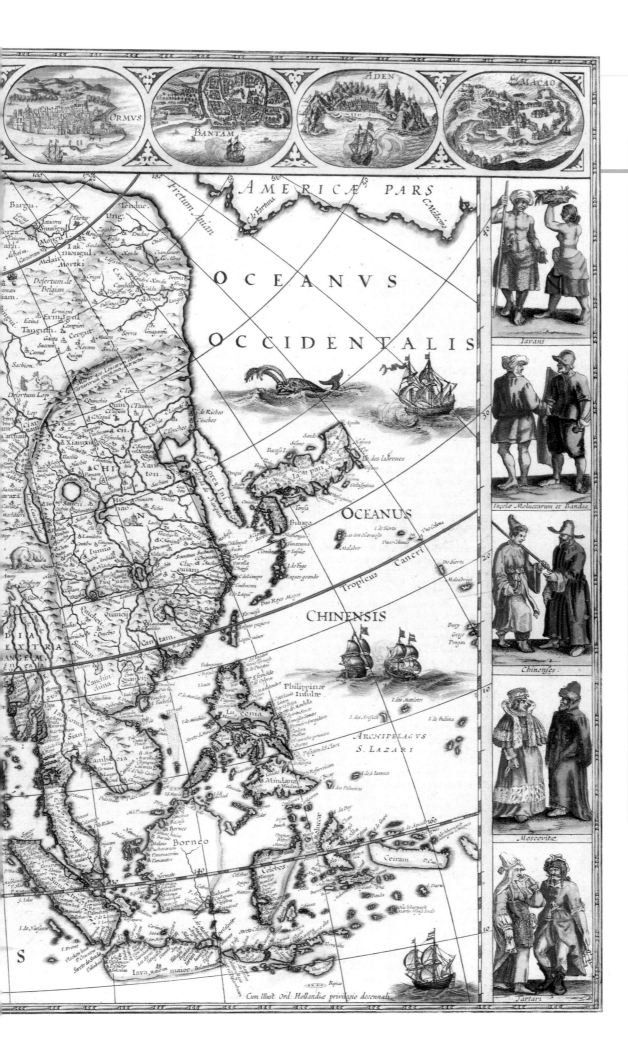

 ORMVS

BANTAM

ADEN

MACAO

AMERICÆ PARS

OCEANVS

OCCIDENTALIS

OCEANVS

Tropicus Cancri

CHINENSIS

Philippinæ Insulæ

ARCHIPELAGVS
S. LAZARI

Borneo

Ceiram

Celebes

Iava maior

Cum Illust: Ord: Hollandiæ privilegio decennali.

Iavani

Incole Moluccarum et Bandae

Chinenses .

Moscovitæ

Tartari

47 Asia, Willem Blaeu, Netherlands, c. 1635

This map, technically termed a *carte-a-figures* (literally, "map with pictures"), was the work of the great cartographer Willem Blaeu, the founder of the Blaeu dynasty of mapmakers who dominated Dutch cartography for more than 40 years in the mid-17th century. The top border shows nine of the main cities and trading settlements of Asia—Candy, Calicut, Goa, Damascus, Jersualem, Hormuz, Bantam, Aden, and Macao—while the side borders show five panels depicting pairs of costumed figures representing the inhabitants of different Asian countries and regions. Blaeu was in a good position to compile the map, as in 1633 he had been appointed hydrographer of the Dutch East India Company, so gaining access to all the geographic information the company had amassed. Willem's sons continued the tradition of fine cartography that their father had established. From 1662, Joan Blaeu published his celebrated *Atlas Major*. Containing 600 maps covering the known world, it was often given luxurious presentation bindings and was a traditional gift presented by the United Republic to royal personages.

Hendrick Christiansen, and twelve merchants presented a petition to the Dutch States General to give them the territorial and trading concessions they needed. It was agreed that the land between New France and Virginia would become the New Netherlands, based on what the States General termed "a Figurative Map hereto annexed" to the charter confirming the grant. This map was christened the Block Map. It was the first to show Manhattan as a separate island and the first to use the words "New Netherlands." It set a pattern for the future, especially when the newly formed Dutch West India Company started taking an interest in the area.

Many more maps of the New Netherlands followed, including ones by Johannes Jansson and Nicholaus Visscher, who improved on Jansson's original. Visscher's 1655 map was the first to mark Nieuw Amsterdam (later New York), which had been built on Manhattan Island on land Pieter Minuit, the Dutch governor of the New Netherlands, had purchased from the local Native Americans for some 60 guilders worth of trade goods. From the European point of view, it was undoubtedly one of the best commercial deals in all of history.

The Dutch in the east

In the first half of the 17th century, the Dutch took over from the Portuguese and Spanish as the most active explorers of the Pacific. In 1618, the Dutch East India Company set up a hydrographical office at its base in Batavia (present-day Jakarta), the capital of the Dutch East Indies, to help with the explorations. In 1642, Anthony van Diemen, Governor-General of the Dutch East Indies, gave the Dutch navigator Abel Tasman (c. 1603–59) the job of "finding the remaining unknown part of the terrestrial globe." In particular, Tasman was set the task of finding the fabled Terra Australis Incognita, the great southern continent that many people of the time believed stretched across the Pacific. Establishing whether or not it existed was a riddle that the Dutch were determined to solve: it was thought likely that Willem Janz, another Dutch sea captain, had made at least an initial sighting of part of it as far back as 1606.

Accompanied by the pilot Frans Jacobszoon Visscher, who had mapped out three possible routes for the voyage, Tasman set sail. First, they visited Mauritius, and then went south. He named his initial landfall Van Diemen's Land—it was later rechristened Tasmania in his honor. As he continued eastward, he made another discovery: New Zealand. He then turned for home, visiting the Tongan islands and Fiji along the way. Two years later, on his second voyage, he landed in Australia itself, christening it Nieuw Holland, a name that was to stick for the next 150 years. After these two voyages Dutch mapmakers were able to show the north, south, and west coasts of Australia as well as incomplete outlines of Tasmania and New Zealand.

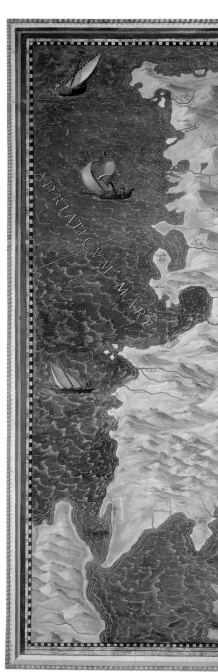

NATIONAL AND REGIONAL MAPS

Map-consciousness spread throughout Europe and reached Russia in the time of Peter the Great (1682–1725). This great age of European discovery and the birth of nationalism went almost hand in hand, as feudalism declined and nation states started to emerge. It was over the same period that rulers, and the bureaucrats who were being increasingly employed to run their governments, started to see how maps could be used to govern more effectively. The first stirrings of such an awareness of the potential use of maps occurred in Italy as early as the mid-15th

48 Sicily, Ignazio Danti, Vatican, c. 1582

As well as advising Pope Gregory XIII on reforming the calendar, Danti, professor of mathematics at the university of Bologna, supervised the creation of the Galleria delle Carte Geografiche in the Vatican. Originally, the intention was to show just the papal territories, but the project eventually expanded to cover the whole of Italy. The maps, which long predated a united Italy, are an expression of papal power and ambition.

century. Pope Pius II (1405–64), described by one contemporary chronicler as "an excellent geographer," read Ptolemy's *Geography* and wrote his own detailed commentary on the importance of the maths of latitude and longitude, while Pope Julius II (1443–1513) commissioned a series of painted wall maps for the Vatican's Loggia del Cosmografia. Later, Pope Gregory XIII (1502–85) brought the scholar Ignazio Danti (1536–86) to Rome to advise on the reform of the calendar, and he stayed to work on a companion set of wall maps in a new Galleria Geografia. The mapping offered not only a "true representation of the Papal states" but covered the whole of an as yet un-united Italy.

Venice, Naples, and Florence were not far behind Rome. As early as 1460, the Council of Ten, the ruling body of Venice, commissioned maps of the territories the Venetians controlled around Padua, Brescia, and Verona, while, late in the 16th century, the kingdom of Naples was extensively mapped by Mario Cartaro (1540–1614), surveyor to the Neapolitan court from 1583 to 1594. He produced a map of the kingdom as a whole, followed by 12 maps that showed each province in detail. At about the same time, Grand Duke Ferdinand I of Tuscany (1544–1604) was displaying his passion for cartography by filling one entire wing of the Uffizi Palace, his residence in Florence, with maps and scientific instruments. Since there was as yet no notion of a united Italy, it is not surprising that an atlas mapping the entire peninsula took time to emerge. This was the work of Giovanni Antonio Magini (1555–1620), who finally managed to publish *Italia* in the year of his death.

The Carta Marina

One of the most curious maps of the entire Renaissance was created by a Swedish Catholic cleric called Olaus Magnus (1490–1557). Despite being created in Venice in 1539 and dedicated to that city's "Most Honorable Lord and Patriarch," who paid for its compilation, it was not a map of any part of Italy at all. The map, which he grandly titled A Marine Map and Description of the Northern Lands and their Marvels, was a richly illustrated portrayal of his Scandinavian homelands, featuring sea monsters galore. Magnus was opposed to the Reformation and was determined to demonstrate to the southern Europeans that Scandinavia was a part of the world that should be saved for Catholicism.

To compile the map, Magnus drew on the maps of the region that had been created to illustrate Ptolemy's *Geography* (Magnus had been inspired to make his map because he considered these to be inadequate). He used many sources, including a collection of mariner's charts he had carefully assembled, and also included observations he had made on his travels. Though the result was still far from geographically accurate, it was the first detailed map of the Nordic countries.

Maps and the Hapsburgs

In the Holy Roman Empire, Charles V (1500–58) was quick to recognize what an important role maps could play in helping him to establish an efficient civil administration in the vast territories he controlled. In Spain, the emperor encouraged Fernando Colón (1488–1539), third son of Christopher Columbus, to start compiling his *Description of the Geography of Spain*, although work on it stopped in 1523. Further north, he commissioned extensive land surveys covering the whole of the Low Countries. He used such surveys as a basis for reassessing levels of taxation.

Philip II of Spain (1527–98), Charles' son and successor, was just as much an advocate of maps and mapping as his father. In the 1560s, he commissioned an extremely ambitious survey of Spain under the direction of Pedro de Esquivel (d. 1575) and completed by Juan López de

49 Carta Marina, Olaus Magnus, Venice, 1539
While living in Italy, exiled priest turned cartographer Olaus Magnus created this vibrant map of his Scandinavian homelands. It covers an area stretching from the North Atlantic to western Russia and from northern Germany to the Arctic Circle. It is a strikingly accurate rendition for the period, although it contains numerous fanciful elements. The sea off the coast of Scotland, for instance, is boiling with dragons and other sea monsters, many of which are busily preying on passing shipping.

50 Cornwall, *Maps of the Counties of England and Wales*, Christopher Saxton, England, 1579

From the early 1570s to 1578, surveyor Christopher Saxton carried out his pioneering survey of England and Wales, consisting of 34 maps. This map of Cornwall is more or less accurate, with the exception of the north coast of the Land's End peninsula, which is shown running east and west, rather than northeast and southwest.

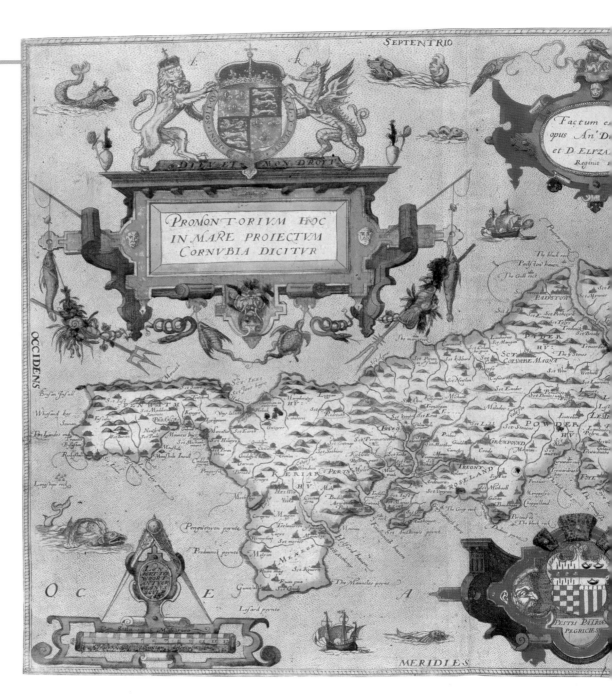

Velasco and João Baptista Lavanha, the compilers of the so-called *Escorial Atlas*. The key map of the atlas showed the entire Iberian Peninsula—Spain and Portugal were united in 1580—and there were 20 maps of its constituent regions.

French national mapping

The story of French national mapping starts with Catherine de Medici (1519–89). The widow of Henri II and effective power behind the French throne for nearly half a century, Catherine, as a Florentine, was well aware of maps and the ways in which they could be used. In about 1560, she charged Nicholas de Nicolay (1517–83), who had been appointed Géographe

du Roi by her late husband in 1552, with the task to "draw up and set out in volumes the maps and geographical descriptions of all the provinces in this kingdom." Nicolay made a start, but in six years he managed to map only three provinces in central France. Whether the delay was caused by the civil unrest that culminated in the drawn-out Wars of Religion between the French Catholics and the Protestant Huguenots is hard to say. Whatever the reason, the project was abandoned.

It was not until 1594 that Maurice Bougereau published his *Le Théatre Françoys*, the first ever atlas of France. However, the volume was made up of maps from different sources. Hence, there was no uniformity of scale, while the coverage was incomplete and the accuracy variable. This, however, did not stop Jean Le Clerk from reissuing it—albeit in an improved form—as *La Théatre Geographique de la France* in 1619.

In 1632, Melchior Tavernier (c. 1564–1644) issued the first map of what he termed the post roads of France. It is interesting to note that, although some of the routes extended as far as Brussels, Basel, and Turin, the ones stretching into Spain, France's greatest political opponent at the time, were deliberately cut short at the Pyrenees, the Franco-Spanish frontier. There matters more or less rested until 1661, when Jean Baptiste Colbert (1619–83) became chief finance minister to Louis XIV (1639–1715) and announced his decision to base his financial and economic policies on a new, exhaustive cartographic survey of the realm.

In response to Colbert's demands, Nicholas Sanson and subsequent French mapmakers developed what would now be termed a base map of the country: a cartographic template on which all sorts of things could be plotted, such as financial districts, religious jurisdictions, river systems, and so on. This concept was to prove extremely influential for mapmakers all around the world.

Mapping the British Isles

Although Elizabeth I was not a map enthusiast, William Cecil, Lord Burghley, who was one of her chief ministers and advisers from 1558 until his death forty years later, and Sir Francis Walsingham, Secretary of State from 1573 to 1590, both used maps extensively. With their interest, English cartography flourished, and surveys began to be carried out from the early 1570s onward, when Christopher Saxton (c. 1542–1606) began to compile his county maps of England and Wales.

Although he has been claimed to be the father of English cartography, not much is known about Saxton's life and career. It seems likely, however, that he learned the basic skills of surveying and mapmaking from John Rudd, a Yorkshire vicar and map enthusiast, who employed him as an apprentice in around 1570. Probably through Rudd's connections,

Saxton was backed by Lord Burghley and Thomas Seckford, a treasury official, to survey the English and Welsh counties, and by 1574 he had his first county map ready for the printers. It depicted Norfolk. By 1578, he had successfully mapped the counties of England and Wales, and the following year an atlas containing all of his maps was published.

Since all the counties had to fit onto a sheet of the atlas, and in some cases more than one county was shown on the same sheet, the maps were by no means uniform in scale: Yorkshire, for instance, was depicted at a smaller scale than any other county. Nor was there any indication of longitude or latitude. The coverage of some of the other counties was sketchy, while no roads were shown at all.

51 London to Bristol, *Britannia*, John Ogilby, England, 1675 John Ogilby created the first English road atlas in 1675. It consisted of 100 maps covering the main coaching routes in England and Wales presented in strip form, with up to seven strips on a single page. All the maps were drawn to a uniform scale of one inch to a mile. This map shows the route from London via Bath to Bristol. The work was an immediate commercial success with four editions appearing within two years of the initial publication.

Saxton's work held the field unchallenged until 1611, when John Speed (1522–1629) produced his own atlas, *The Theatre of the Empire of Great Britain*, which accompanied his *History of Great Britain*. The plates were the basis for subsequent publications and atlases up to the mid-18th century. Speed's county maps featured detailed inset town plans, which formed the first truly comprehensive set of English and Welsh town plans. His map of Wiltshire, for instance, had a detailed plan of Salisbury in the top left corner and a picture of Stonehenge in the top right. Although Speed drew heavily on Saxton and the work of other earlier cartographers, such as John Norden (1548–1625), compiler of the *Speculum Britanniae* (Mirror of Britain) in 1596, many of the town plans were surveyed, as he wrote, during "my owne travels through every province of England and Wales." Speed sent his drafts to be engraved by the great Dutch mapmaker Jodocus Hondius in the Netherlands, and the plates were returned to London for printing.

During his long and colorful life, John Ogilby (1600–76) had many careers, staring off as a dancing master, courtier, and theater-owner and ending up as King's Cosmographer and Geographic Printer. His *Britannia*, a compendium of road maps of England and Wales, was one of the first national road atlases to be produced in Western Europe. It was also the first English atlas to be prepared using a uniform scale, in this case, one inch to a mile. It contained maps of 73 major roads and crossroads presented as continuous parallel strips. Ogilby claimed that he had surveyed 26,000 miles of roads in the course of preparing the atlas, but only about 7,500 miles of them were actually shown in print.

Nevertheless, the result was unquestionably a landmark so far as the story of the mapping of England and Wales was concerned. Maps published after *Britannia* incorporated at least some of the information Ogilby gave. This was by no means confined to roads, because the maps charted other topographical features, such as the towns the roads passed through, rivers, mountains, and bridges. It was little wonder that Ogilby's contemporaries and later mapmakers hastened to pirate his work.

As Britain grew into a colonial power, cartography was employed to express the might of the growing empire. In 1676, Speed's *Theatre of the Empire of Great Britain* was republished, with

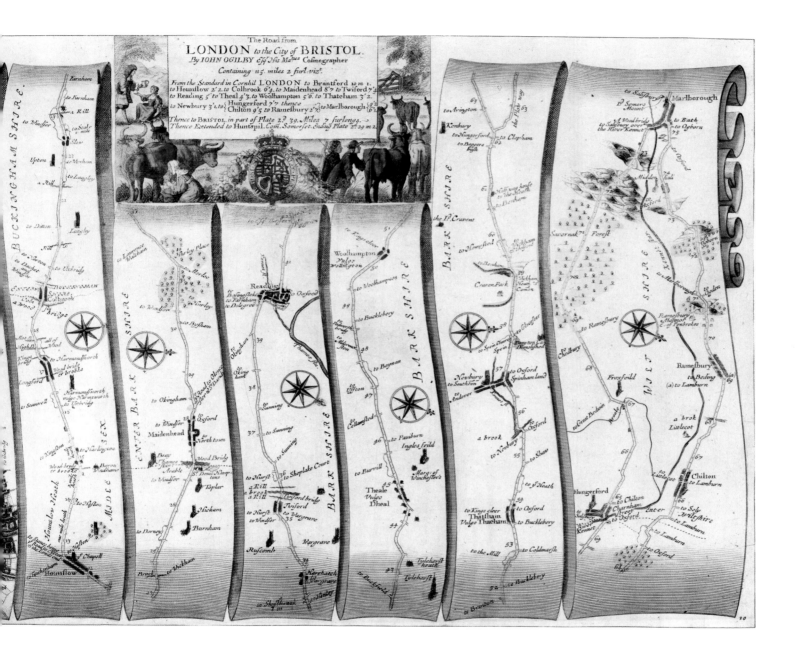

the addition of maps of "His Majesty's Dominions" in New England, New York, the Carolinas, Florida, Virginia, Jamaica, and Barbados.

Estate mapping

It was in England, from the 1570s onward, that an entirely new form of cartography emerged with the introduction of estate maps. They were intended to be a permanent record of a particular landowner's holdings and a tribute to that landowner's power and prestige. They were often drawn for estate management, to show boundaries, or when land changed hands. Estate maps had to be drawn at a scale large enough to allow individual fields and buildings to be shown. They were produced until the mid-19th century, when large-scale Ordnance Survey maps could be used for a similar purpose.

52 Venice, Jacopo de' Barbari, Venice, 1500

This panoramic view of Venice as seen from the southwest was created by painter and printmaker Jacopo de' Barbari. The precision and accuracy of its detail set a new standard for city mapping. The exhaustive survey of the city on which it was based probably took teams of surveyors two years to complete.

What distinguished these maps was that, as their name implies, they normally showed only one single landowner's holdings, leaving other areas blank. If, as was often the case, the holdings were scattered over a wide area—as in William Senior's 1640 survey of the Cavendish estates in Derbyshire and Yorkshire—the result resembled a partly completed patchwork quilt. John Norden had been confronted with a similar problem when he was called upon to map the royal land-holdings in Berkshire, Surrey, and Buckinghamshire in 1607. The result was his *Description of the Honor of Windsor*, which consisted of 17 manuscript maps bound together to make up an atlas. As well as showing the layout of the various royal estates, the scale was large enough for individual deer parks to be mapped. Even the names of the gamekeepers and the number of deer in the parks were noted.

The practice of making estate maps later developed in some other parts of Europe. In the southern part of the Low Countries, for example, Duke Charles de Croy (1539–1613) commissioned a series of estate maps, plans, and views showing the extent of his holdings in what is now Belgium and northern France. In Germany, France, and Italy, use of estate maps did not develop until the 18th century. In the first instance, this may have been one of the results of the upheaval caused by the Thirty Years War (1618–48) and the chaos in the countryside that accompanied it.

CITY AND TOWN MAPS

In the early Renaissance, many Italian rulers, artists, and men of science became passionately engaged in finding new means of recording the world in as realistic and accurate a manner as possible, whether through art or science. As the Italian cities themselves became cultural hot houses, this desire for naturalistic and beautiful reflections of the world, led to a new fashion in mapping. From the late 15th century onward, increasing numbers of city views—striking panoramas that combined miniature, but realistic, depictions of important buildings with an often accurately surveyed ground plan—were produced. Various forms of view developed: for example, the so-called prospect view was a view from the side; a plan or aerial view was seen as if directly from above; and a bird's eye view was a view obliquely from above. Sometimes, several of these views were used in combination in the same illustration.

The map book created for Federigo de Montefeltro of Urbino in around 1470 was typical of this new spirit. It contained no fewer than ten city plans, depicting Milan, Venice, Florence, Rome, Constantinople, Jerusalem, Damascus, Alexandria, Cairo, and Volterra. At around the same time, in Florence, Francesco Rosselli (c. 1445-1527) produced four large woodcuts of Constantinople, Pisa, and Rome as well as of the Tuscan capital itself, while between 1484 and 1487, Pope Innocent VIII commissioned a series of town views for the Vatican in the form of murals that featured Florence, Genoa, Milan, Naples, Rome, and Venice. Six years later, Francesco Gonzaga, Marquis of Mantua, followed the papal example when he commissioned similar murals to decorate his own palace. On this occasion, the chosen cities were Constantinople, Cairo, Florence, Genoa, Rome, and Venice.

Italian plans

Venice was the subject of one of the finest city maps of the Renaissance. In 1500, Jacopo de' Barbari (c. 1440–1515), a Venetian painter and printmaker, was commissioned by Anton Kolb to make a map of the city. Kolb was a businessman from Nuremberg in Germany who had taken up residence there. Barbari created a spectacular bird's eye view looking down on the city from the southwest. His work set new standards in terms of its size, the amount of carefully observed detail it presented, and the care with which the map as a whole was executed. It was printed in Barbari's studio from six

53 Imola, Leonardo da Vinci, Florence, c. 1502

Leonardo's magnificent plan of Imola is believed to be one of the earliest Renaissance geometric town views to have been made. There is some disagreement as to whether he drew the plan from scratch, or altered a map dating from 1473, but the consensus is that he did survey his plan himself, perhaps using the older plan to check his measurements.

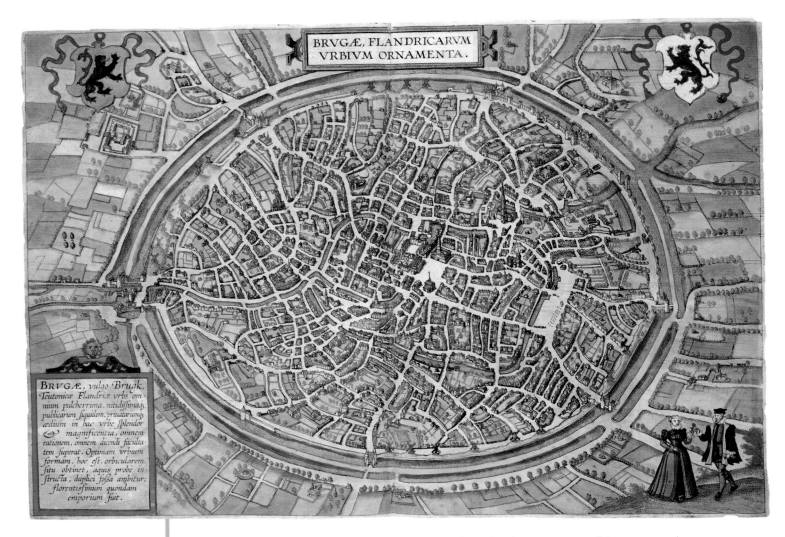

BRVGÆ, FLANDRICARVM
VRBIVM ORNAMENTA.

BRVGÆ, vulgo Brugk,
Teutonicæ Flandriæ vrbs om-
nium pulcherrima, nitidiffimaq,
publicarum fiquidem, priuatarumq;
ædium in hac vrbe fplendor
& magnificentia, omnem
rationem, omnem dicendi faculta-
tem fuperat. Optimam vrbium
formam, hoc eft, orbicularem
fitu obtinet, aquis probe in-
ftructa, duplici foffa ambitur;
florentifimum quondam
emporium fuit.

54 Bruges, Georg Braun and Franz Hogenberg, Germany, 1572

Created by priest and geographer Braun and map engraver Hogenberg, the *Civitates Orbis Terrarum* (1617) was a monumental collection of maps of the cities of the world. It was the first known city atlas, and served as a model for many subsequent atlases. Jacob van Deventer carried out the survey for this map of Bruges.

woodblocks, using the many drawings he had made of various parts of the city over the previous three years as reference.

Three years later, Leonardo da Vinci (1452–1519) produced his celebrated aerial view of the city of Imola at the request of Cesare Borgia, his patron at the time (see p.101). As far as is known, this was the first Renaissance attempt to map a city in planimetric form. Leonardo's later plan of Milan, though simply a sketch, had all the hallmarks of cartographic genius, while the map of Tuscany he created included more than 200 place names. It covered the entire region from the valley of the River Po in the north to Umbria in the southwest.

City atlases

From the 1520s onward, no important German town or city worth its salt was without its own carefully devised visual panorama. One of the most notable of them was the detailed bird's eye plan that was published in 1521 of Augsburg, by far the wealthiest city in the land at the time. The view was created by Hans Weiditz, an artist from Strasbourg, who based his woodcuts on detailed surveys carried out by Jorg Seld, an Augsburg goldsmith and military engineer, between 1514 and 1516. It was the Fuggers, natives of the city and the wealthiest banking family in Europe before their disastrous bankruptcy, who in all probability paid for the plan's compilation and creation.

It was in Germany that the notion of collecting town and city plans together and publishing them in atlas form was first mooted. Georg Braun (1541–1622), a Catholic cleric based in Cologne, had the idea in the 1560s, and in 1572 he and his associate, the engraver Franz Hogenberg (1535–1590), issued the first volume of their mammoth *Civitates Orbis Terrarum* (Cities of the World). By 1617, five further volumes had been produced, featuring a total of about 550 city views in all.

The diversity of the content was extraordinary (Braun and Hogenberg even managed to find and produce a plan of Peking) and the accuracy stunning. The work was a huge success, running through an amazing total of 46 editions in Latin, German, and French. The atlas served as a model for many subsequent city atlases.

The amount of detail the plans presented was variable. Some were more or less straightforward planimetric outlines, as in the case of the plan of Beaumont in southern Belgium, which was based on a drawing by Jacob van Deventer (d. 1575). Others, such as a plan of Cambridge, England, were not drawn to a strict orientation or scale and included scenes of townspeople and animals. While the streets were shown in outline, many of the main buildings, rather than simply having their positions marked, were instead delineated as charming bird's eye views.

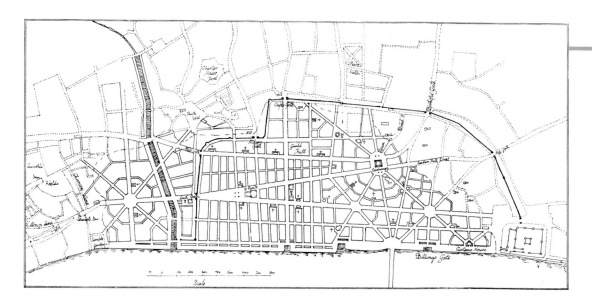

Speculative plans

During the 17th century, maps began to be used in conjunction with city planning. For example, the great architect Sir Christopher Wren (1632–1723) produced an elegant and visionary plan for the rebuilding of London just days after the Great Fire of 1666 had destroyed four-fifths of the city.

Unfortunately, very little of what Wren proposed was built, as it was decided that his scheme was too ambitious and too expensive. It is fascinating to think about what the city might have looked like had Wren's broad tree-lined avenues and open squares replaced the warren of alleys and byways the fire had devastated, and Wren's great "round piazza," his "usefull canal," and "a triumphal arch to the founder of the New City, King Charles II" had actually been erected.

55

Plan for London, Sir Christopher Wren, London, 1666

Within two weeks of the Great Fire of London in 1666, the architect Sir Christopher Wren had produced a detailed plan for the rebuilding of the ravaged city. It was never executed, but in 1749 John Gwynn redrew the plan adding a long commentary on the merits of the proposal. Gwynn also suggested the creation of an official body for "inspecting & condemning old & useless buildings and regulating new ones." Wren's map is a noteworthy example of visionary city planning.

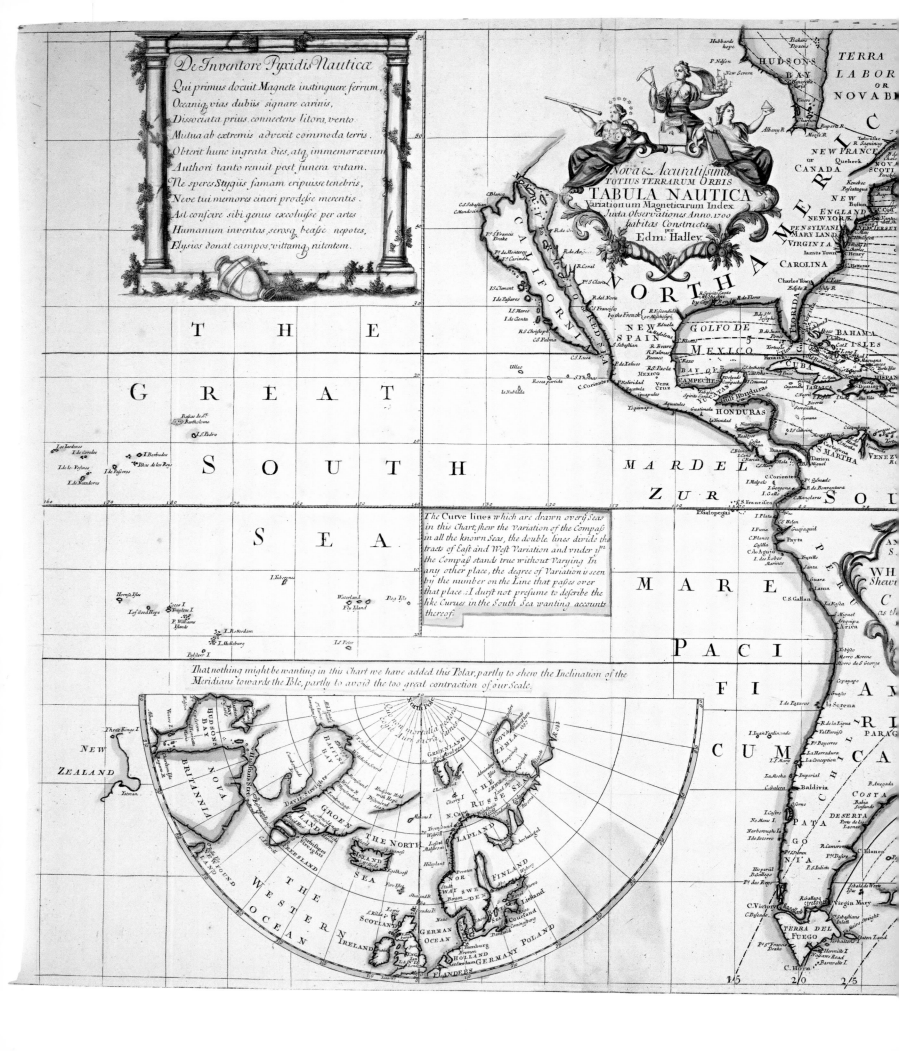

De Inventore Pyxidis Nauticæ

Qui primus docuit Magnete instinguere ferrum,
Oceaniq; vias dubiis signare carinis,
Dissociata prius, connectens litora, vento
Mutua ab extremis advexit commoda terris.
Obterit hunc ingrata dies, atq; immemor ævum
Authori tanto renuit post funera vitam.
Ne speres Stygiis famam eripuisse tenebris,
Neve tui memores cineri prodesse merentis.
Ast conscire sibi genus excoluisse per artes
Humanum inventas, serosq; beasse nepotes,
Elysios donat campos, vittamq; nitentem.

Nova & Accuratissima
TOTIUS TERRARUM ORBIS
TABULA NAUTICA
Variationum Magneticarum Index
Juxta Observationes Anno. 1700
habitas Constructa
per
Edm: Halley

THE GREAT SOUTH SEA

The Curve lines which are drawn overij Seas
in this Chart, shew the Variation of the Compass
in all the known Seas, the double lines divide the
tracts of East and West Variation and vnder ye
the Compass stands true without Varying In
any other place, the degree of Variation is seen
by the number on the Line that passes over
that place. I durst not presume to describe the
like Curues in the South Sea wanting accounts
thereof.

MARE PACIFICUM

That nothing might be wanting in this Chart we have added this Polar, partly to shew the Inclination of the
Meridians towards the Pole, partly to avoid the too great contraction of our Scale.

NORTH AMERICA

NEW FRANCE or CANADA

TERRA LABOR OR NOVAB

HUDSONS BAY

NEW ENGLAND
NEW YORK
PENSYLVANIA
MARY LAND
VIRGINIA
CAROLINA
FLORIDA

NEW SPAIN
GOLFO DE MEXICO
BAY OF CAMPECHE
YUCATAN
HONDURAS

CALIFORNIA

MAR DEL ZUR

BAHAMA ISLES
CUBA
JAMAICA
S. DOMINGO

S. MARTHA
VENEZUELA

PERU

CHILI
PARAGUAY
CUMANA

PATAGONIA

TERRA DEL FUEGO
C. HORN

NEW ZEALAND

NOVA BRITANNIA
HUDSONS BAY
BAFFINS BAY
GROENLAND
ISLAND
THE NORTH SEA
GREENLAND
NOVA ZEMLA

LAPLAND
FINLAND
NORWAY
SWEDEN
RUSSE SEA

SCOTLAND
IRELAND
GERMAN OCEAN
THE WESTERN OCEAN
HOLLAND
GERMANY
POLAND
FLANDERS

THE AGE OF EMPIRE: Pushing the Boundaries

When the British marched out of a besieged Yorktown to surrender to the triumphant American revolutionaries and their French allies in 1781, their bands played a tune called "The World Turned Upside Down." Not only did the tune mark the significance of the actual event, the effective end of the American Revolutionary War, but the tune could be taken as symbolizing the great changes that had been transforming the Western intellectual, scientific, and political world during the 18th century.

In science, figures such as Isaac Newton (1642–1727), Blaise Pascal (1623–62), and René Descartes (1596–1650) had laid the basis for what has been described as a scientific revolution. In thought, men such as John Locke (1632–1704), Voltaire (1694–1778), Jean Jacques Rousseau (1712–78), Adam Smith (1723–90), and Immanuel Kant (1724–1804) in Europe, and Thomas Paine (1737–1809), Thomas Jefferson (1743–1826), and Benjamin Franklin (1706–90) in America had led the way in the open questioning of a host of long-cherished beliefs. What linked all these men was their conviction that any social, political, or scientific issue could be resolved through the judicious application of rational principles. The 18th century is often described as the Enlightenment, or the "Age of Reason," a phrase coined by Thomas Paine in 1795. Cartography was no exception in being deeply affected by the new, challenging intellectual climate.

The 18th and 19th centuries also saw the inexorable rise of the West to dominate much of the world. The Industrial Revolution and the invention of steam power, alongside the demands of trade, were among the most important catalysts. Industrialization also meant the development of new communication routes, first by canal and then by railroad. The routes needed to be surveyed—in France, for example, all railroads built from 1842 onward had to conform to a national plan—and the canals and railroads themselves mapped comprehensively. As far as the latter were concerned, maps were produced by the companies who built and operated the lines and by established map publishers—Rand McNally in the USA was a prime example—all eager to capitalize on a new, expanding market for transport information.

56
Global Magnetic Survey Chart, Edmond Halley, England, 1702
British Astronomer Royal Edmond Halley produced this spectacular chart largely as a result of the observations he made during his voyages on the Royal Navy ship *Paramore*. His mission was to investigate the variation between true north and magnetic north and the possibility of using the difference as a means of calculating longitude accurately at sea. Although the problem of longitude was not actually solved until 1761, Halley's chart helped to make navigation less problematic for subsequent seafarers. Halley failed in his second task: locating the "Coast of Terra Incognita."

57 France,
Jacques Cassini,
France, 1744
Jacques Cassini
followed in the footsteps of his
father, Jean-Dominique, to
create the first map of France to
be based on the triangulation of
the country that the family had
initiated in the 1670s. Aided by
his son, César-François Cassini
de Thury, Cassini finally
completed the 18-sheet
mapping project in 1744. It was
the first accurate cartographic
survey of an entire nation.
Three years later, César-François
embarked on an even more
detailed 175-sheet mapping
program, which was not to be
completed until the 1790s.

In politics, nationalism and imperialism became the two dominant forces as the 19th century
progressed—and cartography responded to meet the needs of the Age of Empire. This led to
increasing government involvement in the creation of what might best be termed administrative
cartography. An early example of this was Napoleon's mapping of France's 30,000 *communes* to
eradicate the administrative geography of the *ancien regime*. The development of national
education systems led to an increasing call for maps in schools and universities, so that, in the
Western world, Victorian children grew up seeing maps displayed on the walls of their classrooms.
Naturally enough, too, the imperial powers were quick to map their new overseas territories.

Cartography was to be equally deeply affected by technological change, as new printing
methods—notably the invention of lithography by Munich printer Alois Senefelder in 1798—

transformed what was rapidly becoming a mapmaking industry by cutting costs, reducing production times, and allowing mapmakers to display new types of data effectively. The coming of the ability to print cheaply in color was also to transform the visual appearance of maps, while the drive to explore meant that more and more geographical information could be reliably depicted. By the end of the 19th century, traditional blank spaces signifying the unknown were rapidly being filled.

As the number and range of maps increased, and reached an ever widening audience, so the traditional ways of making them changed, with the emergence of teams of specialists, from field surveyors, draftsmen, and what were termed "terrain artists" to photographers, engravers, and lithographers. The world of cartography—and the ways in which mapping could affect and respond to the wider world—was never to be the same again.

FRENCH CONTRIBUTIONS

Many cartographic historians hold that truly scientific cartography was born in France during the reign of Louis XIV (1643–1715). In 1666, Jean Baptiste Colbert (1619–1683), Louis XIV's finance minister, brought about the foundation of the Académie Royal des Sciences. Himself an amateur scientist, Colbert was determined to make his country as pre-eminent in science as it was in the arts and war, and the Académie was to be the means of achieving that ambitious aim.

Colbert could count on full backing from his royal master for his plans, for Louis XIV, in common with other "enlightened" European monarchs of the time, saw the supporting of such projects as much as a means of self-glorification as leading to the advancement of knowledge. He not only authorized Colbert to launch a recruiting drive to enlist the services of leading savants from across Europe, but also allowed him to guarantee the members of the new body financial rewards—pensions, as they were termed at the time—that were unprecedented in their generosity. From the start, more accurate and more scientific mapmaking was a priority. Louis himself declared that one of the new institution's primary tasks was to "correct and improve" maps and sailing charts.

Triangulation

One of the cartographic developments in which the French were to lead the way was the adoption of triangulation as a much more accurate basis on which to conduct map surveys. The theory behind the method had been outlined as early as 1533 by the Dutch mathematician Gemma Frisius (1508–58) and was first put to practical use by his fellow Dutchman Willebrord van Roijen Snell (1580–1626), who in 1615 made use of a grid of 33 triangles when he surveyed an 80 square mile (128 square km) area of Holland. France, however, became the first country to employ triangulation on a major scale.

The first step in a triangulation survey is the selection of a baseline of known length. The angles between lines of sight from the baseline's endpoints to a distant landmark are then measured. These distances are then used as new baselines. The process is repeated until the entire area being surveyed is covered by a network, or grid, of triangles. This grid provides the necessary skeleton on which topographic details can be accurately positioned.

Surveying France

In 1668, Jean Picard (1620–82) began a survey that would eventually cover the whole of France. Starting from a baseline stretching 7 miles (11 km) along the road from Paris to Fontainebleau, he became the first man to measure the length of a degree of a meridian, or line of longitude, accurately. Having established this, he went on to triangulate distances to the Channel coast. The survey, which was completed in 1681, revealed that up to that time mapmakers had drawn the country as being bigger than it really was. Louis XIV is said to have commented when presented with the survey's findings that it had "cost me a large part of my state."

Picard was assisted in his survey by the Italian-born astronomer Jean-Dominique Cassini (1625–1712), who had come to Paris in 1669 as a result of Colbert's blandishments. Cassini, who decided to stay in France rather than return to his native country after his appointment as director of the newly founded Paris observatory in 1671, was the patriarch of what was to become probably one of the most remarkable cartographic dynasties of all time. Four generations of the Cassini family were to be involved in compiling what was to become the first accurate topographic survey of France in its entirety. The undertaking—probably the world's first ever national surveying and mapping program—took almost the whole of the 18th century to complete.

The Carte de Cassini

Helped by his son Jacques (1677–1756), Cassini was determined to extend the survey that Picard and he had started south to the Pyrenees and then, as if that goal was not ambitious enough, eventually to cover the whole of the country with a network of triangles. He was not to live to see the Herculean labor through to its triumphant completion. This was to take until 1744, when the first Cassini map of the whole of France was finally completed (see p.106). Consisting of 18 sheets, it can reasonably be claimed to be the first truly accurate cartographic survey of an entire nation. The triangulation on which it was based had involved the measuring, calculating, and plotting of 19 different baselines and some 800 triangles.

The logical step forward was for Jacques and his son César-François Cassini de Thury (1714–84) to use this triangulation as the basis for the creation of further, improved maps. For this, royal financial backing was a prerequisite. Although in 1747 Louis XV (1715–74) gave the project his support, government backing was abruptly withdrawn in 1756. Paradoxically, this was almost immediately after one of the first sheets—there were to be 175 of these in total—had been printed and presented to the king.

The reason for the change of heart was a simple one: it was the cost of the enterprise, which the Cassinis had warned Louis would take an estimated twenty years to complete. France had just embarked on what was to prove a protracted war in Europe and overseas, the Seven Years War, and the royal purse strings were being stretched to their limits. Undeterred, César-François, who had taken over direction of the project from his father, set out to secure a replacement source of funding. He formed a company and sold shares in it to various French notables to raise the finance. This did not come cheap: at its height, there were more than 80 land surveyors, topographers, and cartographers working full time on the project. César-François also did not live to see the task finished. It took until the 1790s for it to be completed, by which time the task of overseeing it had fallen to Jean-Dominique Cassini IV (1748–1845), the great-grandson of the dynasty's founder.

These were troubled times. Following the outbreak of the French Revolution in 1789 and the subsequent overthrow of the Bourbons, the new revolutionary regime made taking control of the Cassini maps a national priority, arguing that they were a prized "National Property." In 1793, the regime dissolved the company and expropriated its assets. Jean-Dominique Cassini was arrested and tried by a revolutionary tribunal, barely escaping with his life. Although, thanks to Napoleon, he was eventually rehabilitated—and his shareholders compensated for their loss—he never regained control of the great map resource that his family had spent so much time and effort creating.

French surveyors carrying out triangulation, Carle Vernet, 1812
This French painting depicts army surveyors using a theodolite. Theodolites, which date from the 16th century, are instruments for measuring both horizontal and vertical angles using a telescope.

NORTH AMERICAN POWER STRUGGLES

While the members of the Cassini family were concentrating on the mapping of France, a wealth of French cartographers were creating ever more detailed and accurate maps of the French possessions overseas and of the wider world.

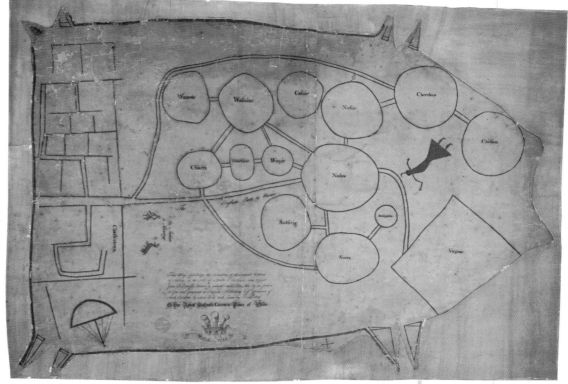

Perhaps the most prominent figure among this new generation of scientific cartographers was Guillaume Delisle (1675–1726), the Premier Geographe du Roi, whose maps of America and Africa were particularly influential.

Delisle's Carte de la Louisiane et du Cours du Mississipi of 1718 was being used by other cartographers as late as 1797 and is thought to have been the oldest map to have been consulted in the planning of the Lewis and Clark expedition that crossed the Rockies to reach the Pacific between 1804 and 1806 (see p.134). The map's remarkable topographical and geographical accuracy made it the template for American mapping for a half-century, but it was also one of the most controversial maps of its day. This was because of the way Delisle chose to treat French territories in North America as opposed to those of Britain. The British colonies along the eastern seaboard were deliberately compressed, while French claims to the territories beyond the Appalachians were exaggerated. The picture the map presents is of a vast river basin, firmly in French possession, and inhabited by a variety of Native American nations. British claims are firmly sidelined.

The Catawba map

It is understandable that the Delisle map pays far more attention to Native Americans nations than it does to the British. Both the French and the British valued the alliances they were at pains to form with the Native Americans, who served both powers as military auxiliaries and scouts. They were also an invaluable source of local knowledge and geographical information. The deerskin map drawn for Francis Nicholson, the governor of South Carolina, in about 1721 by a Catawba Native American (Catawba was the name the British gave to a federation of Native American communities settled between the colonies of Virginia and the Carolinas) is an example of the latter factor in action.

It is thought that the map's creator was probably a chief or headman of the Nasaws, who occupy the central position on the map. He depicted most of what is now the southeastern United States, embracing all or part of the modern states of Virginia, South Carolina, Georgia, Alabama, Mississippi, and Tennessee. The territories of the different nations in the region are indicated by circles. It appears that the main reason behind the map's creation was to demonstrate to the British the important geographical position of the Catawba peoples and that keeping good relations with them was vital for the maintenance of land communications between the British colonies in the region.

58 Catawba Nations, America, 1721 Drawn on deerskin, this map was probably created by a chieftain of the Nasaw group of Native Americans for Francis Nicholson, the British governor of South Carolina. The map was intended to play an important role in cementing relationships in the region. Virginia and the Carolinas, the two most important British colonies in the South, are depicted on the map's margins with the territories of a powerful Native American confederation of tribes separating them. Each circle on the map roughly indicates the position of a tribe or settlement.

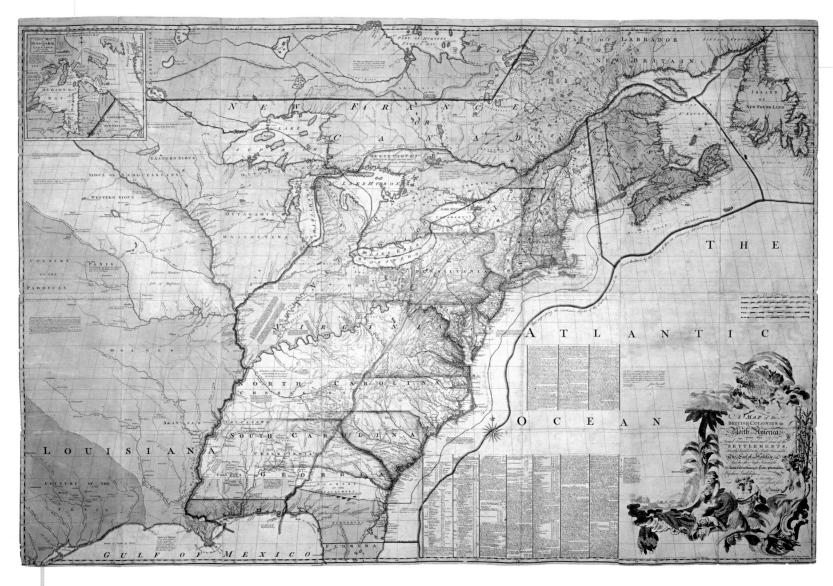

Map wars

Delisle was not the first French cartographer to juggle with geopolitics in his attempt to overstate France's territorial claims and play down those of the British. Nicholas Sanson (1600–70) had used much the same tactic in his famous 1656 map *Le Canada, ou Nouvelle France* (see p.76). What had changed in the interim was the relationship between the two powers. In the time of Oliver Cromwell (1653–8) and Charles II (1660–85), relations between Britain and France had on the whole been friendly. By the early 18th century, however, France was seen as the greatest threat to Britain's burgeoning commercial and imperial interests. The British fought and won a protracted war, known as the War of the Spanish Succession (1701–14), to curb the ambitions of Louis XIV. The two nations were to remain more or less at loggerheads until the defeat of Napoleon at Waterloo in 1815.

It was little wonder, therefore, that the British quickly encouraged their own cartographers to produce maps that would carve up North America more in Britain's favor. Just two years after Delisle's map had been published, map publisher Herman Moll (d. 1737), a German geographer who had arrived in Britain in 1678 and had already published *A New and Exact Map of the Dominions of the King of Great Britain on Ye Continent of North America* in 1715, produced *This Map of North America According to Ye Newest and Most Exact Observations*. It was intended as a complete counterblast to Delisle's claims. In Moll's map, the British colonies in North America stretch from the Carolinas to Newfoundland, while the Labrador Peninsula is firmly labeled "New Britain." Louisiana and New France seem to shrink in size by comparison, while, off the eastern seaboard, the Atlantic Ocean has changed its name: it is now entitled the "Sea of the British Empire."

59
British and French Dominions in North America, John Mitchell, England, 1775

The Mitchell map, also known as the "Red Line" map, was generally regarded as the most accurate and detailed map of North America of its times, although, as was its creator's intention from the start, it stressed British territorial claims over French ones. The red lines on this edition of the map were the work of the negotiators at the Paris peace conference of 1783 that ended the war between Britain and the American colonists. Drawn in red ink by Richard Oswald, the chief British negotiator, they demarcated the newly agreed boundaries between the United States and British-held Canada.

America Septetrionalis

The overriding idea behind this and other British maps of the time was to marginalize existing French holdings on the continent and to push the boundaries of the British colonies outward. This was the motive behind the decision by the Lords Commissioners of Trade and Plantations to commission the British mapmaker Henry Popple (d. 1743) to produce a large-scale map of the British Empire in America, plus "the French and Spanish settlements adjacent hereto." The result was Popple's *America Septetrionalis*, published in 1733. Consisting of 20 regional maps, each appearing on a separate sheet, plus an index map showing the entire continent including the Caribbean, it is believed to be one of the largest maps to be produced in the 18th century. In compiling it, Popple drew heavily on the information he had gathered first-hand during the time he spent working in the North American colonies, and he also used the Delisle map for reference.

The Popple map was highly regarded by many. John Adams, later to become the second President of the United States, wrote in 1776 that "Popples [sic] map is the largest I ever saw and the most distinct." Unfortunately, the map, although undeniably packed with authentic detail, did not please the men who had commissioned it. The colonial rivalry between Britain and France was spiraling: in 1756 the two countries began the Seven Years War as a result of this rivalry. It was felt that the map did not go far enough in providing cartographic support to bolster the British position. In 1755, the British delegates at a conference held to try to delineate British and French territorial claims formally disowned it. Since it would be incautious to attack it politically, instead they disparaged it as inaccurate. Their comment was that "it has ever been thought in Great Britain to be a very incorrect map, and has never in any negotiation between the two Crowns been appealed to by Great Britain as being correct, or as a map of any authority."

A new map of North America that appeared in the same year that the Popple map was disowned doubtless met with a better response among British officialdom. Issued under the auspices of the aptly named Society of Anti-Gallicans, and compiled by William Herbert and Robert Sayer, the map's geography was entirely biased in Britain's favor. The British territories, which were vastly exaggerated in size and extent, were picked out in bold color, while what remained of the French holdings on the continent—these were disparagingly referred to on the map as "French possessions and encroachments"—were left colorless.

The Mitchell map

Another map, also first published in 1755, was to prove highly influential. In 1750, the Earl of Halifax, then the President of the British Board of Trade, had asked John Mitchell (1711–68), a physician, botanist, and cartographer, to prepare a major new map of the British colonies in North America. It was to take Mitchell five years to complete the task as he corrected and recorrected his drafts in pursuit of his goal of accuracy, particularly as far as latitude and longitude were concerned. He was determined not to repeat the errors he believed had been made by earlier mapmakers, whom he felt had "grossly misrepresented" the geography of the continent.

Like Popple, Mitchell had personal American experience to draw on: he had emigrated to Virginia in 1720, returning to Britain 26 years later. He was given full access to the various surveys, maps, and reports in the possession of the Lords Commissioners and held in their archives. The Commissioners also ordered the governors of the colonies to furnish Mitchell with new maps and detailed descriptions of the territories they controlled.

Mitchell's Map of the British and French Dominions in North America as it finally emerged is generally held to be one of the most significant in early American history. Its coverage embraces almost half the North American continent, stretching south from Hudson's Bay to central Florida

60

Mason-Dixon Line, Charles Mason and Jeremiah Dixon, America, 1768

When they were employed by the Calverts of Maryland and the Penns of Pennsylvania to settle the disputed boundary between their land holdings between 1765 and 1768, British surveyors Charles Mason and Jeremiah Dixon could never have expected that their survey would come to feature so prominently in later American history. Their boundary line is seen here in an 1850 map. From 1820, when slavery was banned north of 36°30' latitude by the Missouri Compromise, the term "Mason-Dixon line" came in general usage to mean the boundary between free states and slave states.

and west from the Atlantic seaboard to the present-day Kansas-Missouri border. What is immediately noticeable is that, to meet the needs of the Lords Commissioners—Mitchell was fully in sympathy with their fear of "the designs of the French in all parts of America"—the boundaries of Virginia, the Carolinas, and Georgia were extended across the Mississippi and west into what looks like infinity. The British possessions were also highlighted in strong, solid colors to reinforce the notion of ownership.

The amount of detail that Mitchell managed to include on his map was impressive, particularly so far as the Native American inhabitants of the territories he was mapping were concerned. He mapped three categories of Native American settlements—towns and forts, villages, and deserted villages—which were accompanied by notes and commentaries. The latter broadened out into general descriptions, such as "A Fine level and Fertile Country of great Extent, by Accounts of the Indians and our People," especially when it came to the interior and western regions of the map.

What made the Mitchell map such a widely respected document was its undoubted locational accuracy. Between its first publication in 1755 and the end of the 18th century, a total of 21 different editions of the map were published in four languages—notably in French and German—and it became the standard base for all subsequent cartographic representations of the continent. It was regarded highly enough to be used as the reference for settling the new boundaries between Britain and France in the Treaty of Paris that ended the Seven Years War in 1763. Defeat in this war cost the French Canada and forced them to give up all their claims to the continental interior. The 1775 edition of the map was employed to help to settle the boundaries between the newly independent United States and its neighbors in a second Treaty of Paris that brought an end to the American Revolutionary War in 1783.

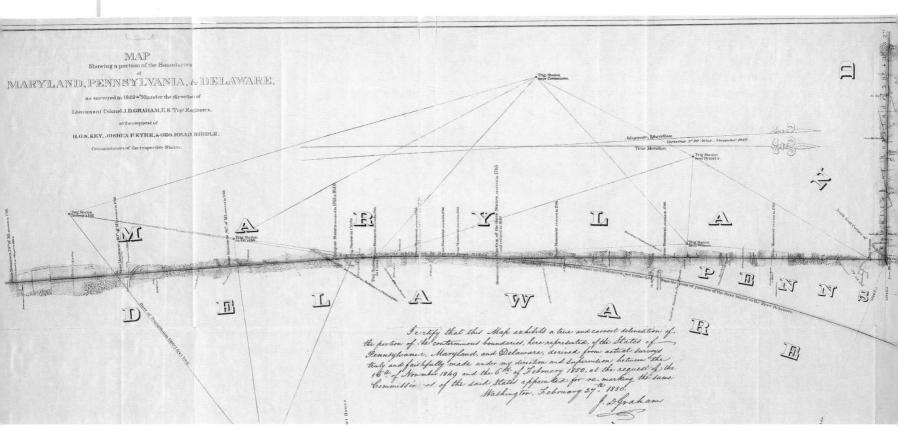

The Mason-Dixon Line

In 1763 a more localized boundary dispute gave rise to another highly influential map when British surveyors Charles Mason (1730–86) and Jeremiah Dixon (1733–79) were asked to settle a dispute between the property owners of Maryland and Pennsylvania. The survey took until 1768 to complete, largely because work on it had to be suspended for a crucial period in the face of opposition from Native American peoples.

The Mason-Dixon line, as it became known, initially ran for 244 miles (393 km) west from the Delaware border. In 1773, it was extended to the western limits of Maryland and, in 1779, was extended again to delineate the eastern boundary of Pennsylvania with Virginia (present-day Western Virginia). In the period up to the outbreak of the American Civil War in 1861, the term "Mason-Dixon line" popularly came to stand for the border separating slave states from free ones. Even today, it is still used to distinguish the South from the North.

BRITISH DEVELOPMENTS

In the 18th century, Britain started to emerge as a world leader in the field of mapmaking, but developments were often reactive, seeming to be sparked by responses to specific events. Following the near success of the 1745 Jacobite rising, the British government commissioned William Roy (1726–90) to produce a new and detailed survey of Scotland at a scale of one inch to 1,000 yards for use by British forces in the event of any future rebellion. The founding of what was to become the Ordnance Survey in 1791 was a similarly pragmatic response to another perceived military threat, this time from revolutionary France.

Work on what at the time was called the Trigonometrical Survey began in 1791. Building on the triangulation that had been started in 1784 to link the Royal Observatory, just outside London, and the Paris Observatory, parties of surveyors were soon hard at work. By 1795, a double chain of triangles had been completed between London and Land's End at the western tip of Cornwall and, in 1801—the year in which the Ordnance Survey was officially named—the first map at a scale of one inch to the mile was published. It was of Kent, one of the natural targets for any French invasion force (see p.114). The surveyors then moved on to the southwest, completing their work there by about 1810 and then slowly but steadily progressing to cover the rest of the country. As far as England and Wales were concerned, the surveying process was finally completed in 1874, though black-and-white maps of the whole of both countries—with the exception of the Isle of Man—were being published by the start of the 1870s. From 1897 onward, color was to take over from black-and-white. Nor was the work of the Ordnance Survey confined to national mapping. In 1855, work started on the mapping of the nation's cities. By 1892, all British cities with more than 4,000 inhabitants had been mapped at a scale of 1:500.

The Ordnance Survey started work in Ireland and Scotland in 1825 and 1837 respectively, although the Irish survey was not the first detailed one to be carried out in that country. Some surveying had taken place in the time of Oliver Cromwell, when the Puritan regime confiscated more than half the land in the country as part of the savage reprisals taken against the Irish for their support of the Stuart cause. Sir William Petty (1623–1687) was charged with organizing the so-called Down Survey of the confiscated areas. He got the job by convincing Cromwell and his government colleagues that he could complete the survey in an amazing 13 months. Rather than pay for proper surveyors, Petty used demobilized—hence inexpensive to employ—soldiers from Cromwell's army to conduct the fieldwork. The results were then sent back to Dublin for skilled cartographers to plot and chart. Such economies helped Petty to make his fortune out of the project.

MASON AND DIXON LAYING OUT THE BOUNDARY LINE

Mason and Dixon Laying Out the Boundary Line, illustration from *The Leading Facts of American History* by D.H. Montgomery, 1890 Mason and Dixon's boundary line entered both the popular imagination and the American lexicon.

61

County of Kent with Part of the County of Essex, Ordnance Survey, England, 1801

This map was created by the "surveying draftsmen" of the Ordnance Survey on the basis of a trigonometrical survey carried out in 1799 by William Mudge, a captain in the Royal Artillery. Consisting of four sheets, it was the first map that the Ordnance Survey produced, before continuing to map the whole of England and Wales by 1874. The map was initially published privately by William Faden, Geographer to George III, and not by the Ordnance Survey itself.

Enclosure maps

The large-scale Ordnance Survey mapping was particularly useful for property and taxation purposes. With no further need for it, the British tradition of national and regional mapping, which had taken off in Tudor times, went into decline, but there was still a call for estate and enclosure maps. Mostly dating from the mid-18th century to the mid-19th century, the latter were instruments of land reorganization that both reflected and consolidated the power of those who commissioned them. In Britain as elsewhere, the process of enclosure, which involved the fencing or hedging in of vast tracts of farmland and common land, was initially sparked off by the impact of the Agricultural Revolution. Prior to the Agricultural Revolution, farmers subsistence-cropped strips of land in fields held in common. The developments in agricultural mechanization during the 18th century required large, enclosed fields in order to be workable. Enclosure continued well into the

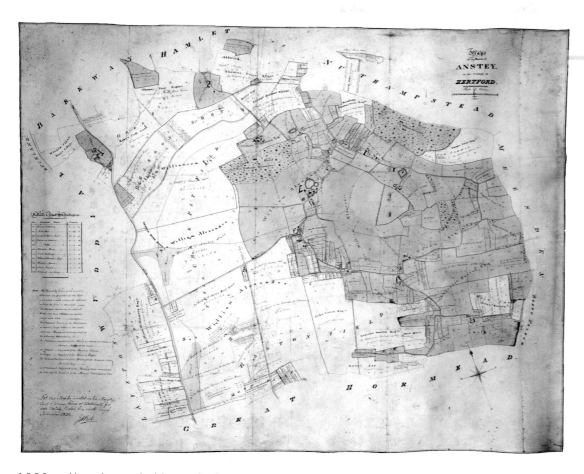

62

Enclosure Map,
England, 1829
In England, enclosure—
the means by which
open fields and commons became
private plots of land—reached its
height in the early 19th century. From
the 1780s, maps such as this became a
regular feature of the process. They
showed who was allotted land, who
bought land, what the landscape
looked like, and how the land was
affected by the enclosure. Here, the
ownership of individual areas of land
around the village of Anstey in
Hertfordshire is indicated, while the
"ancient footways" that were to be
blocked off are also marked.

1800s, although it probably reached its peak in Britain during the Napoleonic Wars (1803–1815), when it is estimated that, in some areas, nearly a third of the land was enclosed. The maps that resulted fell into two categories. Sometimes, although not very often, the surveyors would make a kind of pre-survey, showing what the land looked like before a specific enclosure got under way. In the main, however, they confined themselves to mapping the enclosed areas after the event.

Enclosure was by no means a purely British phenomenon. From the mid-18th century onward, hardly an enclosure anywhere in Europe was conducted without its being mapped. Thousands of maps resulted. Although individually they were only of local relevance, taken as a body, they epitomized a socio-economic change that was to transform the way of life—and the landscape—across the entire continent. In country after country—notably in France, where, from the mid-18th century, many nobles took their hereditary lands in hand with a view to maximizing their productivity, Italy, and Germany—landowners came to recognize that such maps were an essential part of effective estate management. Indeed, they enabled landowners to manage their holdings so efficiently that the surveyors making them were sometimes physically threatened by tenant farmers and agricultural laborers. The farmers foresaw, often correctly, that the arrival of a surveying party would lead to a consequent demand for an increased rent.

Enclosure maps varied greatly in terms of their quality, the scale they employed, and the amount of detail they showed, but the best of them offer an invaluable insight into the rural landscape of the time. The field-by-field surveys of titheable land that were instituted by the Tithe Commutations Act of 1836, although intended to be extremely detailed, similarly vary in quality from parish to parish, largely as a result of local opposition to their cost.

**Edmond Halley by Sir
Godfrey Kneller
(1646–1723)**
In 1698 Halley received a
commission as captain of
HMS *Paramore* to make
observations on the
conditions of terrestrial
magnetism. This task he
accomplished in a two-
year voyage, completing a
chart which extended from
52°N to 52°S.

EXPLORING THE PACIFIC

The English astronomer and mapmaker Edmond Halley (1656–1742) is today best known for his 1705 identification of the celebrated comet that now bears his name. However, his maps included a star chart, a map showing the directions of the world's trade winds, and what was possibly his most influential production: a global map detailing the variation in degrees between magnetic and geographic north (see p.104). In 1698, Halley had been commissioned by the British Admiralty to try to find a way of determining longitude more accurately on board ships, and this map, which was published in 1702, was the result. Halley based it partly on the observations he had made on an expedition to the southern part of the Indian Ocean and also on the charts that had been produced by various navigators working for the Dutch East India Company. Although the problem of longitude was not actually solved until 1761, Halley's chart certainly helped to make navigation less problematic for subsequent seafarers.

Great seafarers

The mapping of the Pacific, which, up to the mid-18th century, had been full of error, was one of the great adventure stories of the century, in which explorers, mariners, and mapmakers were all to play a part. The Portuguese explorer Ferdinand Magellan (c. 1480–1521) had been one of the first sailors to attempt to chart part of the Pacific while he was making the first circumnavigation of the world in the service of Spain. Sir Francis Drake (c. 1540–96), the great Elizabethan seadog, had followed in his footsteps. The Dutch sailor Abel Tasman (c. 1603–59) had become the first European to reach what later turned out to be Australia and to visit Tasmania, New Zealand, Tonga, and Fiji.

Now it was the turn of the British and the French to take up the torch. From the 1760s, the achievements of men such as Louis-Antoine de Bougainville (1729–81), the commander of the first French expedition to the South Seas in 1764; Jean-François Galoup La Pérouse (1741–88), explorer of the northwestern Pacific; and, above all, James Cook (1728–79), whose three expeditions from 1768 onward produced the first accurate nautical charts of the Pacific Ocean, transformed European knowledge and understanding of the entire region.

In little more than 50 years, this immense area, covering more than one-third of the Earth's surface, was no longer the mystery it had been for so long: it had been thoroughly charted and mapped. In the process, some hoary geographical traditions, such as the belief in the existence of a Terra Australis, the great southern continent, had been swept away once and for all.

Chronometers, triggerfish, and stick charts

One advance that unquestionably eased the task of the new generation of European Pacific explorers—and the mapmakers who accompanied them on their voyages—was the discovery of a reliable means of determining longitude at sea accurately. Up to the 1760s, the problem seemed insoluble. Then, in 1761, Lincolnshire clockmaker John Harrison, in pursuit of the substantial reward of £20,000 (£2 million today) that the British Board of Longitude had offered for a solution to the problem, perfected his celebrated ship's chronometer—what he termed a "marine timekeeper." The accuracy of Harrison's chronometer proved to be amazing. It meant that it was now possible not just to measure local time, but also to establish the time at another fixed reference point. For James Cook's second Pacific voyage, for instance, the chronometers he took with him were calibrated at Plymouth just before he sailed. With this information, it was now a relatively simple matter to determine any longitude mathematically and accurately.

Of course, the Europeans were not the only or the first peoples to have ventured across the unknown expanses of the Pacific. Long before their arrival on the scene, the peopling of what Westerners term Oceania—by convention, the area is usually subdivided into Micronesia, Melanesia, and Polynesia—had begun. Two theories have been advanced to explain these migrations. The first holds that they were the result of mishaps or accidents, the second that they were the fruit of deliberate, planned explorations. Both theories may well be correct.

What we know for sure is that the Pacific islanders possessed considerable navigation abilities. One of the most important things they did was to make accurate records of the time it took them to canoe from island to island. This was the only gauge of distance that concerned these nomadic seafarers. There is also no doubt that they developed their own forms of cartography. Throughout Oceania, for instance, it seems likely to have been common practice to tie knots in a cord as a means of recording and memorizing geographical features.

Different peoples developed their own individual methods and traditions. One of the ways in which the Caroline Islanders mapped their archipelago, for example, was through the use of a cartographic device known as the "great triggerfish." These are large-scale maps that got their name from their visual resemblance to the triggerfish. They show five places, four of them forming a diamond to represent the head, tail, and dorsal and ventral fins of the fish. The head is always the eastern point and the tail the western one, but the dorsal and ventral fins can be either

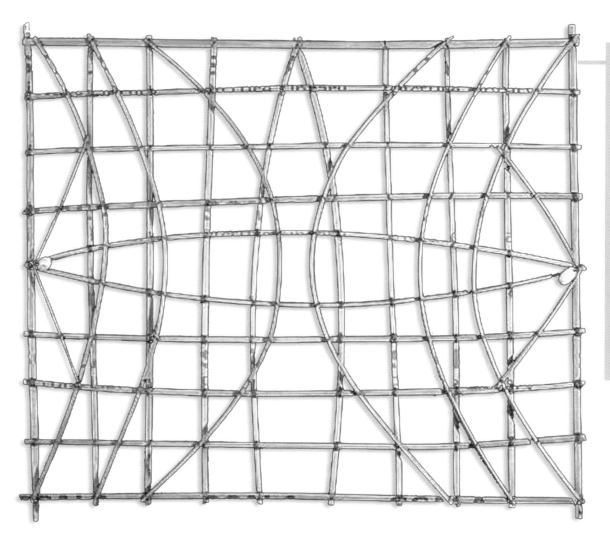

63

Maritime Stick Chart, Marshall Islands, 20th century
The Marshallese, the inhabitants of two parallel island chains in the heart of the Pacific, created these so-called stick charts as a means of recording ocean currents and wave patterns around the islands and atolls in their part of the ocean. The ways in which the sticks were tied together represented the varying patterns, while seashells were employed to signify islands and atolls. They were navigational aids rather than tools, and were rarely, if ever, taken on voyages.

northern and southern or southern and northern points respectively. The fifth place in the center of the diamond represents the fish's spine.

The Marshall Islanders also developed their own form of mapping, creating three types of what are known as "stick charts"—the *mattong*, the *rebbelib*, and the *meddo*—as instructional navigational aids (see p.117). Both the *rebbelib* and the *meddo* were employed to record information about such phenomena as wave patterns, the direction of ocean swells, and how far away a particular island could be seen from a canoe. Seashells, or the intersection of the wooden strips that were used to make the charts symbolized the islands themselves.

Cook's three voyages

Though it may seem invidious to ascribe particular credit to any one of the Western explorers who ventured into the unknown reaches of the Pacific, there is little doubt that the figure who most revolutionized its mapping was James Cook. In the course of his three voyages to the South Seas, he discovered and charted more of the Pacific Basin than any other explorer before or since.

Cook's first voyage, which lasted from 1768 to 1771, took him to Tahiti to observe a transit of Venus. He sailed on from there to map successfully the coasts of New Zealand and eastern Australia, where he landed at Botany Bay. He originally named the place Stingray Harbour after the plenitude of such fish in its waters, but later changed it to the name we know today because of exotic plant specimens that were discovered there by Joseph Banks (1743–1820), the young official botanist to the expedition. He sailed on from there as far as the Great Barrier Reef, where his ship, the *Endeavour*, struck razor-sharp coral and had to be hastily beached to be repaired.

Once temporary repairs had been completed, Cook headed for the Dutch East India Company's base in Batavia, where the *Endeavour* could be fully overhauled. There, he entrusted copies of the three most important charts he had drawn—the first was of what he termed the "South Sea," the second was of New Zealand, and the third showed the area of Australia that he had christened New South Wales—to the officers of a returning Dutch merchant fleet, who had agreed to forward the charts to the Royal Society in London. News of Cook's most significant cartographical discoveries on the voyage thus arrived in Britain before he did.

Cook's second voyage, which started in 1772 and ended in 1775, took him back to New Zealand, Tahiti, Tonga and its surrounding islands, Easter Island, the Marquesas, the New Hebrides, and New Caledonia. It ranks high among the greatest exploratory voyages of all time. During it, Cook sailed further south than anyone else had managed before him, crossing the Antarctic Circle and continuing until a barrier of pack ice forced him to turn back. The reluctant conclusion Cook reached was that Terra Australis, the legendary great temperate southern continent for which he and many of his fellow explorers had been searching, simply did not exist.

Cook's third and final voyage (1776–9) is best remembered for his discovery of Hawaii, where he perished during a clash with some Hawaiians. It is clear, though, that the discovery was the consequence of a chance sighting as Cook sailed from the Society Islands toward the northwest Pacific coast of America, rather than as the result of a planned search. Cook's orders from the British Admiralty were to find a way around North America through the fabled Northwest Passage, which many believed linked the Atlantic with the Pacific. He hoped that he would meet up with another British expedition that was simultaneously trying to locate the passage from the Atlantic side of the continent. Cook sailed far enough to enter the Arctic Sea through the Bering Strait, but failed to find any trace of the passage he was seeking. Nevertheless, he managed to fix the longitude of the coast, thereby proving that the North American continent was far wider than had previously been thought. This was to have great repercussions for future American mapmaking.

Cook's ship landing in Hawaii, engraving by Meno Haas, 1803
Cook's most notable accomplishments were the British discovery of the east coast of Australia, the European discovery of the Hawaiian Islands, and the first circumnavigation and mapping of Newfoundland and New Zealand.

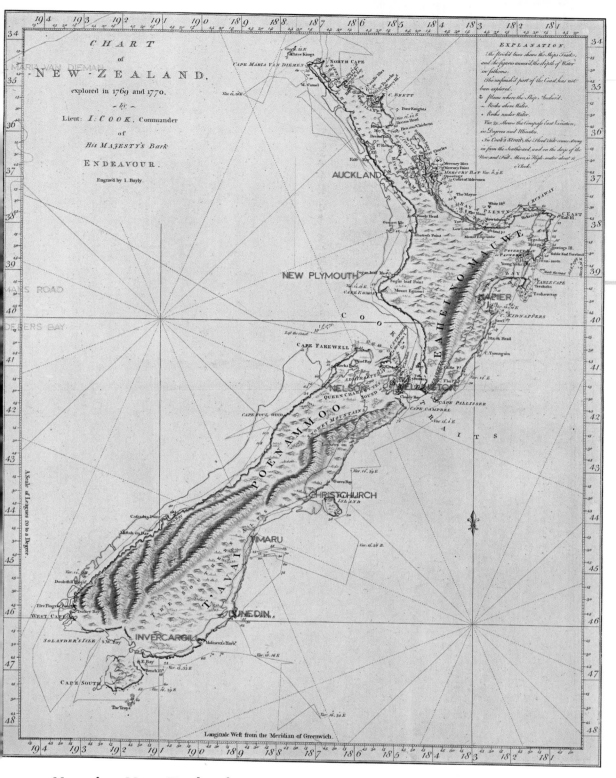

CHART
of
NEW-ZEALAND,
explored in 1769 and 1770,
– by –
Lieut: I: COOK, Commander
of
His MAJESTY's Bark
ENDEAVOUR.

Engrav'd by I. Bayly.

EXPLANATION.

64 New Zealand, James Cook, England, 1773 British explorer James Cook mapped the coastline of New Zealand during his first voyage to the Pacific between 1768 and 1771. His survey showed that New Zealand consisted of two main islands—the North Island and the South Island—separated by the strait that Cook discovered and which is now named after him. His achievement was all the more remarkable since at times his ship was forced out to sea by bad weather. This meant that he was forced to rely on guesswork for the mapping of parts of the coast.

Mapping New Zealand

Cook had established his reputation as a cartographer long before setting out on his Pacific Odyssey, particularly through his mapping of the St. Lawrence and the coast of the Carolinas in North America. In the Pacific his skills were to be tested to the full. He started his remarkable survey of New Zealand immediately upon his arrival there in 1769, his aim being to establish whether or not this strange new land was part of a southern continent. He named his initial landfall Poverty Bay "because it afforded us not one thing we wanted." Nicholas Young, the ship's boy who made the first sighting, was rewarded with a gallon of rum, and the bay's southwestern point, Young Nick's Head, was named after him.

Sailing south from Poverty Bay and then north, Cook charted inlets, bays, capes, and promontories as he progressed along the coast. Some of the names he gave to these features were

65

River Mouth Map, Dr. David Malangi, Australia, 1983

David Malangi (1927–99), an Australian Aboriginal artist, created this map-painting on bark depicting the mouth of the River Glyde in Central Arnhem Land, Northern Territory. Its content reflects Malangi's understanding of his world and his place within it. This picture contains geographical references—the river flows into the sea at the bottom of the painting—and thematic ones. Many contemporary Aboriginal artists continue to work in traditional formats, keeping alive community traditions both artistic and social.

obvious enough—Hawkes Bay, for instance, was named in honor of Lord Hawke (1705–81), the First Lord of the Admiralty—but others were more individual. Cape Kidnappers, for example, was christened after a Maori attempt to abduct a Tahitian boy who had enlisted in the crew.

Cook fairly quickly realized that what he was surveying was an island, a supposition he verified by climbing a hill to survey from a height. The question was whether the land he could see lying farther to the south was a second island or part of the great southern continent he was seeking. Eventually, hard work, careful observation, and information received from a friendly Maori elder, established that the former, not the latter, was the case. Cook had established that New Zealand consisted of a northern and southern island.

As far as North Island was concerned, Cook believed that the charts he had made were "pretty accurate," but he was slightly less confident about his mapping of South Island. In his Journal, he wrote:

> **"** *The Season of the year and circumstance of the Voyage would not permit me to spend so much time about this island as I had done at the other and the blowing weather we frequently met with made it both dangerous and difficult to keep upon the Coast.* **"**

Despite these reservations, Cook had nevertheless completed his main mission successfully. Thanks to the detailed chart he had made of the coastline of both of its two major islands, he had transformed the accepted cartographic picture of New Zealand in less than six months.

Aboriginal maps

Cook noted that the Aboriginal peoples he met around Botany Bay were fairly indifferent to their European visitors. He was surprised, for instance, that they showed no interest at all in acquiring iron. Like many Europeans since and the Dutch before him, Cook showed no real understanding of Aboriginal culture, although it would have been hard for him to have done so given the short-lived nature of his encounters. Until relatively recently, Aboriginal cartography was widely misunderstood and unappreciated. Since Aboriginal maps lack orientation and regular scale, the reaction of most Westerners was to dismiss them completely. Now, however, the definition of what is and what is not a map has been rewritten into something far less blinkered. The topographical representations produced by Aboriginal artists over countless generations are maps, regardless of what form they take.

There are two main types of Australian Aboriginal maps: the bark paintings of the tropical northern and northeastern parts of Australia and the so-called "dot painting" maps of the Outback, especially of the area around Alice Springs in the Northern Territory. Both types share a common origin and purpose. They are part of a broader system of meaning, in which there is no clear division between now and then, the present and the past. Such maps are far more than topographical representations: they do not simply show where things are, but also celebrate their deeper meaning as part of the Aboriginal Dreamtime, when the world was created.

The supernatural beings of the Dreamtime inhabit the physical features of the land, shaping them and giving them life. A failure to understand and communicate with the spirits could be fatal. Australian Aboriginal mapping is a means of passing on Dreamtime myths and the spiritual and practical wisdom that they carry. This teaching could be the safest route from one waterhole to another or the handing down of social customs and laws.

Surveying the Australian coast

The Aborigines were to be left in peace for quite a few years after Cook's visitation, since it was not until the late 1780s that the British found a use for the territories he had claimed for them. Following the loss of the American colonies, the British decided that a new home for transported convicts had to be found, so the Botany Bay colony—the first of several—was established in 1788.

In making its decision, the government was deeply swayed by the evidence that Cook's botanist Joseph Banks gave to a House of Commons committee investigating the issue. Banks was confident in his belief that such a colony could be made to pay for itself, while, so far as the unfortunate convicts were concerned, he opined that "escape would be very difficult, as the country was far distant from any part of the globe inhabited by Europeans."

Settlement gradually progressed, even though cartographic knowledge of Australia was relatively slow to expand. Until 1798, for instance, it was commonly and erroneously believed that, rather than being an island, Tasmania was linked to the Australian mainland. This belief was disproved only when George Bass (1771–c. 1803) and Matthew Flinders (1774–1814) sailed through the strait that separated the two.

Between 1801 and 1803, Flinders went on to survey the coasts of southern and western Australia, but the publication of the chart he made was delayed. On his return voyage to Britain, major damage to his ship forced Flinders to put into the French-held island of Mauritius. This was at the height of the Napoleonic Wars and he was promptly interned as a spy. He did not succeed in regaining his freedom until 1810, by which time a rival French expedition had compiled its own maps of the area. These were issued in 1811, three years before Flinders managed to get his own chart into print. Flinders and Nicholas-Thomas Baudin (1754–1803), his French counterpart, had met during the course of their explorations at aptly named Encounter Bay. As finally compiled by Louis de Freycinet (1779–1842), the French maps were resolutely Napoleonic. For example, the name Bonaparte was given to the modern-day Spencer Gulf and Josephine to Gulf St. Vincent, while the eastern segment of the southern Australian coast was christened Terre Napoleon in honor of the French emperor.

Into the Australian interior

Outside the scattered British settlements on and close to the Australian coast, the interior of the country remained largely unknown and unmapped, although there were various theories about what it might contain. One of the most popular was that it was home to a great inland sea. In 1827, for example, a map drawn for the publication *Friends of Australia* showed this sea and "the supposed entrance of the Great River" flowing from it to the northwestern coast.

The map of southern Australia produced in 1854 by James Wyld (1812–87), the official Geographer to Queen Victoria, as part of his *The New General Atlas of Modern Geography*, was far more informed. It was probably the most accurate survey of the region to that date. Its production might well have been sparked off by the great Australian gold rush, which began in 1851, since known gold deposits were marked on it.

Wyld benefited in the compilation of his map from the various explorative surveys that had taken place in the preceding years. In 1813, a way had been finally found across the Blue Mountains. More surveying expeditions were led by John Oxley (c. 1785–c. 1828), who discovered and named the Lachlan and Macquarie Rivers, and Thomas Mitchell (1792–1855), whose four expeditions took him into hitherto unknown areas of southeastern Australia and led to the opening up of new grazing lands in the southern parts of what would become Victoria. Mitchell named these lands "Australia Felix." Other explorers of note included Edward Eyre

Drafting pen, compass dividers, and pocket telescope belonging to Captain Matthew Flinders Flinders was one of the most accomplished navigators and chartmakers of his age. He circumnavigated Australia, identified the effect of iron ships upon compass readings, and wrote the seminal work on Australian exploration, *A Voyage To Terra Australis.*

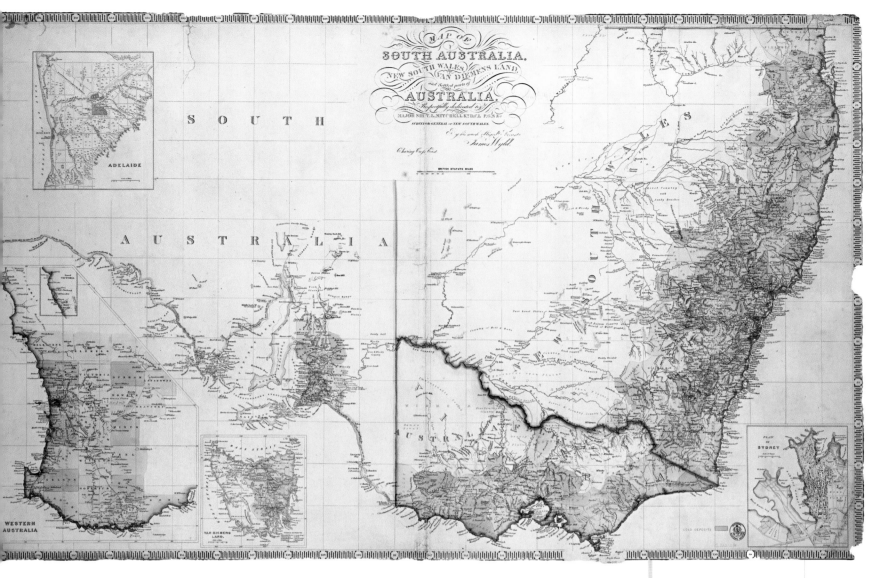

(1815–1901), the first European to travel across southern Australia from east to west, and Charles Sturt (1795–1869), who, by pioneering an overland route to Adelaide from Sydney, added greatly to cartographic knowledge of the interior of Australia.

The quests of three explorers, however, ended in tragedy. The disappearance of Ludwig Leichhardt (1813–48), along with the other members of his expedition, during his attempt to reach Perth from Brisbane in 1848 remains one of the greatest unsolved riddles of Australian cartographic history. What happened to Robert O'Hara Burke (1821–61) and William John Wills (1834–61) is equally tragic, though not so much of a mystery. They succeeded in becoming the first men to cross Australia from south to north, but died from starvation on their return journey.

THE BRITISH IN INDIA

The mapping of India presented British cartographers with a daunting challenge. Up until the middle of the 18th century, British cartographic knowledge of the interior of the subcontinent was limited. Given the ever-expanding role of the East India Company—largely as a result of its defeat of Siraj Ud Daulah, the last independent Nawab of Bengal, at the battle of Plassey in 1757—this was a situation that could not be allowed to continue. There was an urgent need for an accurate survey that would show the company, in clear cartographic terms, exactly what it was it controlled, what it did not, and how best to defend its holdings against possible attack.

66

South Australia, James Wyld, England, 1854

James Wyld, Geographer to Queen Victoria, was one of the most prolific and prestigious mapmakers of mid-Victorian times, mapping everything from the battles of the Crimean War to the main routes of the London and South West Railway. His map of South Australia, which incorporated all of the most recent discoveries, might well have been inspired by the great Australian gold rush of 1851. The insets are of western Australia and Van Diemen's Land, together with city plans of Adelaide and Sydney.

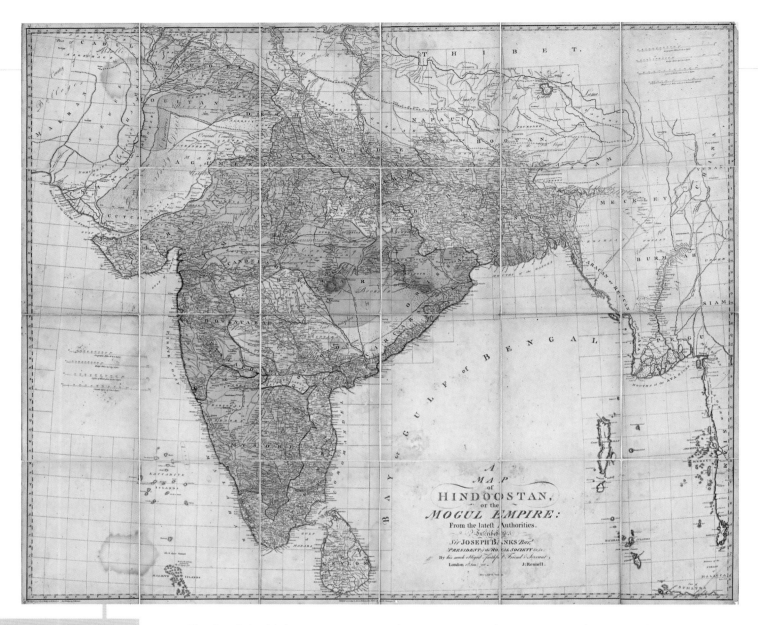

Hindoostan, James Rennell, England, 1788
The first Surveyor-General of Bengal from 1767 to 1777, James Rennell is widely regarded as the father of Indian geography for his pioneering mapping work. In his maps, Rennell used the old Mughal province divisions as they had been promulgated during the reign of the Emperor Akbar (1556–1605). With its lavish hand coloring, this wall map is typical of his passion for detail. It extends from the barely explored "Himmalahs" in the north to Ceylon in the south, and west to the "Great Sandy Desert."

The British had little to go on, since, despite ancient India's major contributions to the development of mathematical science, relatively few geographical maps survived from the pre-European period. It was actually two Englishmen, Thomas Roe (c. 1581–1644) and William Baffin (c. 1584–1622), who compiled the first detailed map of the Mughal Empire in 1619 based on Roe's experiences during a four-year sojourn at the court of the Great Mughal in Agra. Cosmographical maps, on the other hand, were plentiful, stemming from all three of the most important indigenous religions: Buddhism, Hinduism, and Jainism.

The first steps to creating a survey of the subcontinent were taken by James Rennell (1742–1830), who was Surveyor-General of Bengal for a ten-year period from 1767. Ordered by Robert Clive, then Bengal's Governor-General, to "set about forming a general map of Bengal with all expedition," he had begun the task of surveying the River Ganges and the surrounding area in 1766. His method involved measuring distances and directions along the major roads to produce what he called "route surveys," accompanied by charting of the latitudes and longitudes of major features. The result was the *Bengal Atlas*, published in 1779 and based on no fewer than 500 individual surveys, followed by his first Map of Hindoostan, which appeared in 1782.

The Great Trigonometrical Survey

At the turn of the 18th century, the East India Company took the bold decision to embark on the systematic triangulation and mapping of the subcontinent. William Lambton (c. 1756–1823) was entrusted with what was to prove the Herculean labor of establishing a network of geodetically measured triangles stretching from the southern tip of India north to the Himalayas and across the breadth of the subcontinent from Bombay to Calcutta. It was a task that was to take getting on for a century to complete. The process started with the measurement of a baseline between two points—these were usually 7 miles (11 km) apart—with the aid of a chain of precisely known length mounted on wooden trestles. This took several weeks. Then, the angles between either end of the baseline and the sightline to a third point were measured with a theodolite. Where there were no hills handy to establish the lines of sight, flagmen, perched precariously on top of lofty towers of flimsy bamboo scaffolding, substituted. Many of them fell to their deaths.

Another reason for the slowness of the progress was disease: malaria and other killer tropical illnesses struck down entire surveying parties. Nevertheless, by the time of Lambton's death in 1823, the triangulation had got as far as halfway up the spine of India. George Everest (1790–1866), who succeeded Lambton as Superintendent of the Great Trigonometrical Survey, as it had been named in 1817, pressed ahead with the task. By the time Everest retired in 1843, the survey had been extended as far as Dehra Dum in the foothills to the north of Delhi, while arms of triangulation were starting to reach east and west along the Himalayan glacis.

By the 1860s, the survey had embraced much of India's turbulent northwest frontier. The accurate mapping of this area was a priority given the perceived threat to what would poetically be christened the "jewel in the crown" of the British Empire from expansionist Tsarist Russia. The companion topographic survey took from 1825 to 1906 to accomplish. The result was an amazing collection of no fewer than 358 detailed maps, all using a uniform scale of 1 inch to 4 miles.

68 **Burmese-Chinese Frontier, Shan artist, Burma, 1889**
An anonymous Shan artist—the Shan were a tribal people whose territories straddled Burma's border with China in the 19th century—painted this tempera map to help settle a frontier dispute between the British-protected Shan state of Möng Mäo and the Chinese Empire. The Möng territory is colored red and China yellow. The map covers an area of around 47 square miles (75 sq km) along the Nam Mao River. The ovals indicate the locations of hamlets and villages, while the irregular green shape signifies the town of Namkhan.

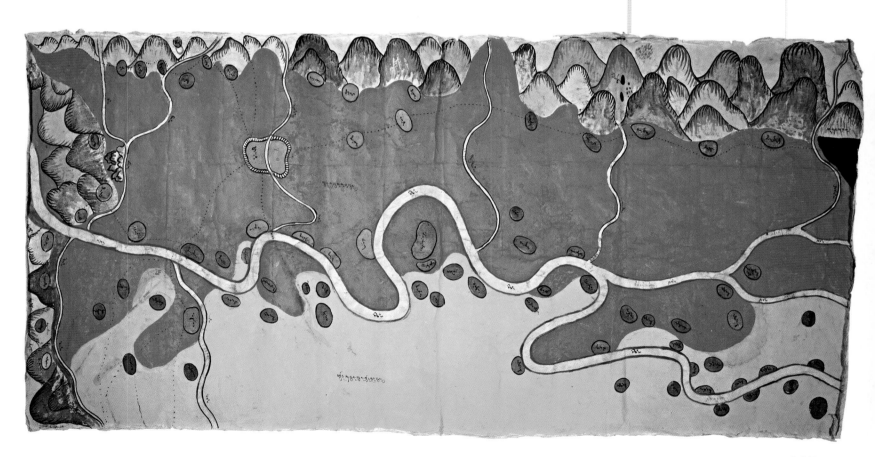

69

World Map, Japan, 1853

This world map, which dates from 1853, was probably the last to be issued before the arrival of Commodore Perry's American military and trade mission put an end to Japan's self-imposed isolation from the outside world. Like various others, this particular map was based on a 17th-century original: the Great Map of Ten Thousand Countries that was created for the Chinese emperor Wanli by the Jesuit missionary Matteo Ricci (1552–1610) during his stay at the imperial court in the early years of the 17th century. Knowledge of the map reached Japan from China during the early Edo period, which lasted from around 1603 to 1867, when the ruling shoguns were overthrown and the emperor reassumed supreme power. Such maps were one of three categories of world map the Japanese produced, the oldest being maps that reflected Buddhism and its traditions, such as the belief that a giant mountain called Mount Sumeru stood at the center of the world. The most modern, dating from mid-Edo times, were influenced by Dutch maps of the period, brought to Japan by merchants. All three different worldviews managed to coexist through the Edo period.

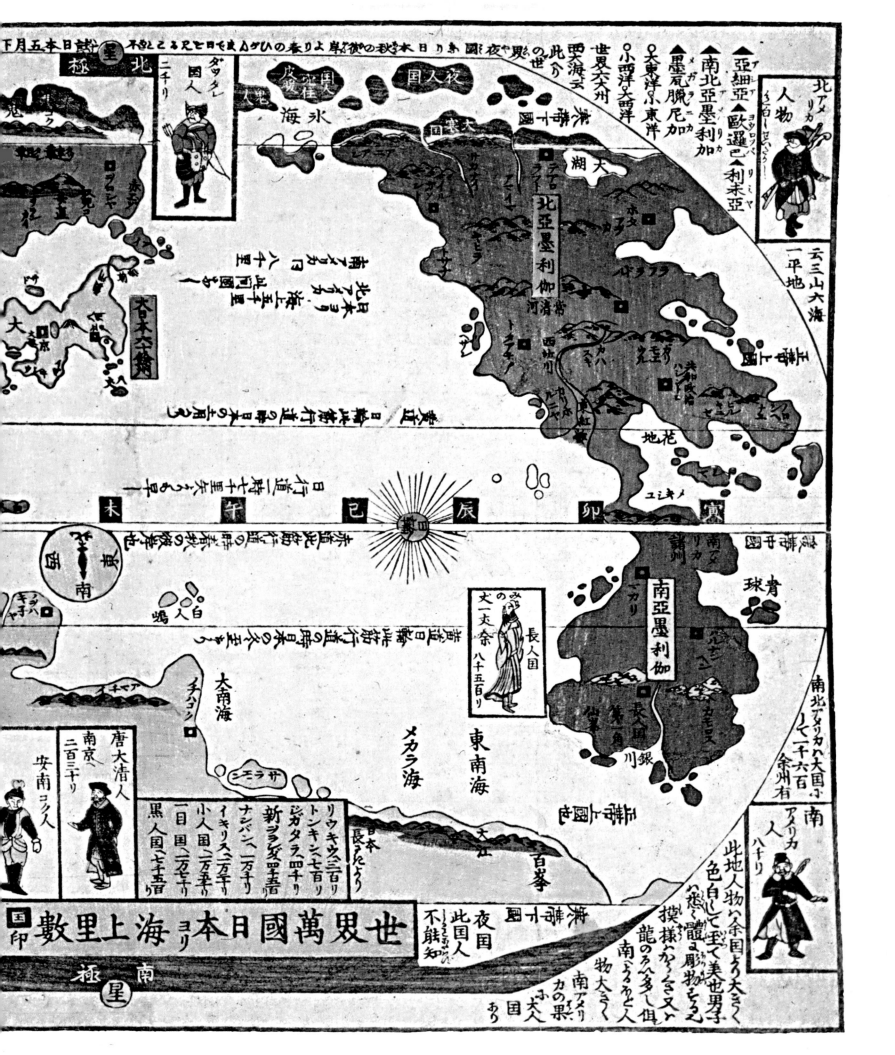

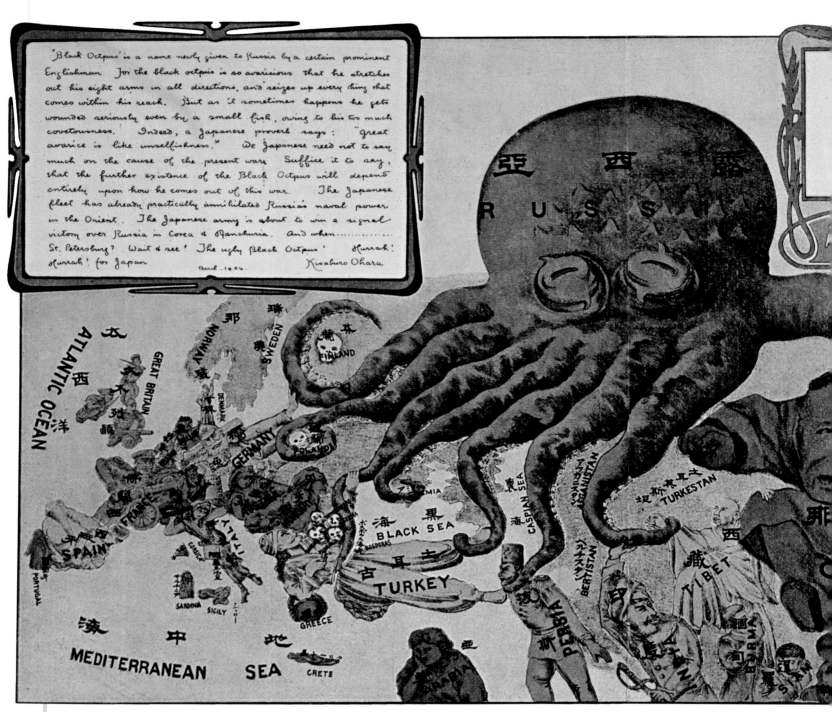

'Black Octpus' is a name newly given to Russia by a certain prominent Englishman. For the black octpus is so avaricious that he stretches out his eight arms in all directions, and seizes up every thing that comes within his reach. But as it sometimes happens he gets wounded seriously even by a small fish, owing to his too much covetousness. Indeed, a Japanese proverb says: "Great avarice is like unselfishness." We Japanese need not to say much on the cause of the present war. Suffice it to say, that the further existence of the Black Octpus will depend entirely upon how he comes out of this war. The Japanese fleet has already practically annihilated Russia's naval power in the Orient. The Japanese army is about to win a signal victory over Russia in Corea & Manchuria. And when............ St. Petersburg? Wait & see! The ugly Black Octpus! Hurrah! Hurrah! for Japan

March. 14.04.   Kisaburo Ohara

The octopus of domination

British determination to resist what were held to be Russia's overweening territorial ambitions was not confined solely to India. The Balkan crisis of 1877, in which the British government sided with the Ottoman Turks against the Russian tsar, led to the publication of a curious Serio-Comic War Map by the noted Victorian graphic artist and caricaturist Fred W. Rose. It illustrated the threat posed to British interests by the Russian octopus in its quest for world domination. Rose continued to produce eye-catching cartographical curiosities for the rest of the century, including what was probably his masterpiece, Angling in Troubled Waters: A Serio-Comic Map of Europe, in 1899.

70 Serio-Comic War Map, Fred W. Rose, England, 1877

Caricaturist Fred W. Rose created the original version of this cartoon map in 1877, when British hostility to Russian territorial ambitions in the Balkans, at the expense of the tottering Ottoman Empire, was at its height. This map, which is annotated in English and Japanese, is a reworking of the same theme by a Japanese artist in March 1904, immediately prior to the outbreak of the Russo-Japanese War. The avaricious octopus's tentacles extend into northern and eastern Europe, the Middle East, and Asia, where China is about to fall victim to their embrace.

That Britain had created a mighty world empire on which "the sun never sets" was in no doubt at all by the 1880s, as the colorful map of it produced in 1886 based, as its legend has it, "on statistical information provided by Captain John Colomb" amply demonstrated (see p.130–1). John Colomb (1838–1909) was a major imperialist thinker of the day, arguing that the navy was the most important frontline component in the scheme of imperial defense he so fervently advocated.

CARVING UP AFRICA

One of the areas of the world of most interest to the European imperial powers of the mid- to late 19th century was the interior of Africa, which remained largely unmapped. The continent's mapping was yet another cartographic challenge to be taken up by a generation of British Victorian explorers. In 1857, for instance, the London-based mapmaker and publisher John Arrowsmith (1790–1873) produced his Map of South Africa showing the "routes of the Reverend Dr. Livingstone between the years 1849 and 1856" (see p.132).

Dr. David Livingstone (1813–73), originally a shopkeeper's son from Blantyre, Scotland, was to become the most celebrated African explorer of Victorian times. He sailed for southern Africa on behalf of the London Missionary Society in 1840. Over the following years, he steadily moved north into central Africa. In 1851 he reached the Zambezi River. From there, he set out on a series of travels through what was then still largely Arab slave-trading country, discovering the Falls of Shangwe along the way. These he renamed the Victoria Falls in honor of the British queen. By the 1870s, he had reached the Nile Basin, where he died of dysentery in 1873. His surveys, drawings, and observations were brought to the coast by the African members of his expedition and transported to Britain.

Three other British explorers contributed to the opening up to Europeans of what was still described as the "Dark Continent." Sir Richard Burton (1821–90) discovered Lake Tanganyika; John Hanning Speke (1827–64) found Lake Victoria; and Sir Samuel Baker (1821–93) discovered Lake Albert. By the time Henry Morton Stanley (1841–1904) published his Through the Dark Continent, with ten maps, in 1878, the blank areas on the African map were gradually being filled.

What followed in the 1880s and 1890s was the infamous scramble for Africa, in which the interior was carved up into European colonies, protectorates, and spheres of influence with no concern at all for the indigenous inhabitants. It seems as if the great powers and their mapmakers were more concerned with creating a neat appearance on paper than in anything else, as on many occasions the new boundaries that were decided on, largely as a result of the Berlin Conference of 1884–5, cut clean through existing tribal lands, territories, and kingdoms.

71 British Empire, John Colomb, England, 1886

Based on statistical information compiled by Sir John Colomb, this map shows the British Empire as it approached its zenith, although, if it had appeared ten years later, much of the African continent would also have been colored pink. The map contrasts British landholdings across the world at the time it was created with those of a century earlier, in the inset. The black lines indicate the main sea routes that held the empire together. The map was produced by the Imperial Federation League, which had been founded under the chairmanship of the Liberal imperialist politician Lord Roseberry two years earlier and soon had branches throughout the empire. The map was published in the weekly newspaper the *Graphic* as a special color supplement to mark Queen Victoria's Golden Jubilee. Such maps impressed on the reader the might and reality of the empire. Colomb was a prominent imperialist thinker, writer, and politician of the day, who was an ardent advocate of an integrated system of imperial defense.

FRATERNITY FEDERATION

MAP OF THE WORLD.
SHOWING THE EXTENT OF THE BRITISH TERRITORIES IN 1786.

WORLD

HUMAN LABOUR

72 Southern Africa, John Arrowsmith, England, 1857

Victorian mapmaker John Arrowsmith published this map to capitalize on the public interest in the adventures of David Livingstone, probably the most celebrated African explorer of the Victorian era. Arrowsmith, who was head of the major map publishing concern founded by Aaron, his father, had published his first map of South Africa in 1834. On this map, Livingstone's route is marked, while the rivers and outlines of lakes delineated with dotted lines were "from Oral information generally."

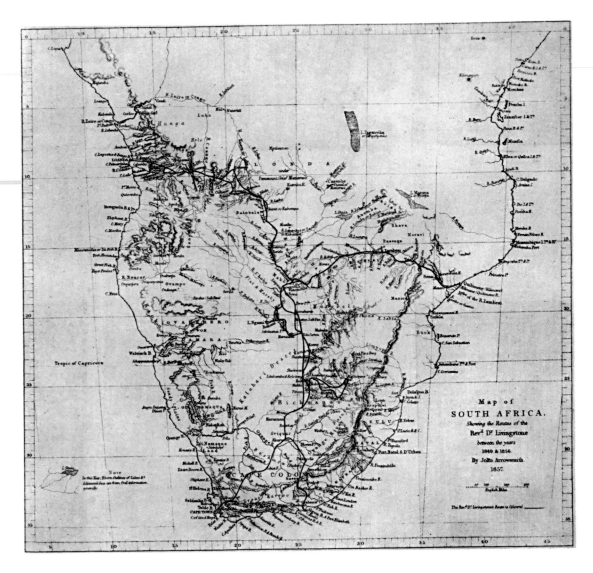

Indigenous African maps

Although Europeans of Victorian times and later did not know it, Africa had a particularly rich tradition of mapmaking. At various times and in various places, African cultures produced cosmographic maps, mnemonic maps to aid in the retelling of origin myths, rock and sand maps, village plans, and maps of kingdoms. In addition, they created maps of rivers and caravan routes—many drawn in the sand—at the request of European explorers.

The kingdom of Bamum in western Cameroon was the scene of one of the most ambitious mapmaking exercises. Led by Sultan Njoya, the Bamum people undertook a major topographical survey of the kingdom, which the king presented to the British in 1916. It took 60 Bamum surveyors two months of solid work to create. Its probable purpose was to consolidate Njoya's regal power: even the small decorative marks on the map appear to have been designed to enhance this image.

MAPPING THE UNITED STATES

Once it had won its independence from Britain in 1783, the newly founded United States began a process of territorial expansion that slowly but surely was to lead to the creation of a mighty nation stretching "from sea to shining sea." During the 19th century, as millions of European migrants flocked into the country in search of freedom and opportunities, wave after wave of settlers moved ever further westward, so by leaps and bounds extending the republic's frontiers.

More territory was acquired in various ways. Louisiana was purchased from France as early as 1803, while in 1848, following the American annexation of Texas and as the result of the defeat of the Mexicans in the war that ensued, the United States gained control of California and much of the present-day American southwest. In 1867, Russia sold Alaska to the United States, while, following a protracted dispute over the border with Canada that at one stage brought the Americans and the British close to open war, Britain agreed to a peaceful settlement of the two nations' territorial claims.

While this expansion was continuing, successive generations of American mapmakers faced the challenge of mapping and charting these vast new territories. The geographer Jedidiah Morse (1761–1826), a friend of both Benjamin Franklin and George Washington, stressed the pressing need for such an effort early on in the life of the infant republic. Morse's *The American Geography*, published in 1789, was enormously popular and influential. As originally published, there were only two maps in the book: one of the Northern states and the other of the Southern ones. Morse was quick to point out in his introduction that, up to that time:

> " *Europeans have been the sole writers of American Geography and have too often suffered fancy to supply the place of facts, and thus have led their readers to errors, while they have proposed to aim at removing their ignorance.* "

The mapping process had started before this, even before the US constitution itself had been formally ratified. Thanks to Thomas Jefferson, who chaired the congressional committee charged with working out the plan, the Continental Congress enacted the 1785 Land Ordnance—or, to give it its full title, Ordinance for Ascertaining the Mode of Disposal of Lands in the Western Territory. The new government needed money and the quickest way of raising it was to sell off the vast territories it now theoretically controlled.

Accordingly, the decision was made to divide the land up into convenient geographical squares, orientated north to south and east to west into what were termed townships and ranges. Such a division, it was thought, would be simple to understand, easy to undertake, and cheap to

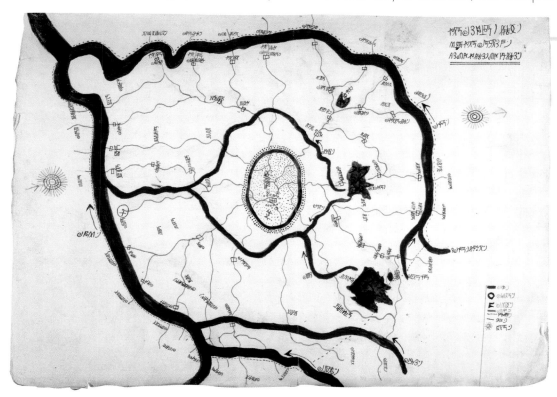

73 **Map of Sultan Njoya's Kingdom, Cameroon, 1916**
Sultan Ibrahim Njoya had his kingdom in western Cameroon surveyed and mapped in the early 20th century, perhaps to impress his subjects and the colonial powers who were active in the area with his importance. The map presents the image of a stable state, effectively masking power struggles and creating a sense of unity. In common with other similar African maps, the process of selection, omission, and positioning of information was clearly influenced by the desire of the mapmakers to influence specific social and political situations.

survey. The Ordnance and subsequent legislation, such as the Homestead Act of 1862, sparked off the greatest transfer of land from public to private ownership in world history—and maps were a vital part of the process, since a map was needed in order to claim title. Almost 23 million acres (92 million hectares) of land were involved.

The Lewis and Clark expedition

The consequent surveying became primarily a local or state responsibility and was often contracted out into private hands. Up until the late 1830s, what federal mapping there was seems to have been executed irregularly, carried out by army surveyors working on an informal, almost ad hoc, basis. However, this did not prevent some remarkable maps from being made. Among the most famous of these maps were the ones created by Meriwether Lewis (1774–1809) and William Clark (1770–1838) on their great expedition up the Missouri River, across the Rocky Mountains, and on into the far west of the continent until they eventually reached the Pacific coast.

The expedition, launched under the auspices of President Thomas Jefferson (1801–09)—Lewis was Jefferson's private secretary and Clark his former commanding officer—started in May 1804. In November 1805 it reached the Pacific, returning to St. Louis in September 1806. Clark was the expedition's cartographer, drawing all but three of the 140 maps that were drafted during its course. He started off by making daily route sketches. Subsequently, once safe in winter quarters, he combined these sketches with information he drew from various Native American maps to which he had access to create composite regional maps, charting the various stages of the expedition's epic journey. The most notable of these Native American maps was probably one prepared by Ac ko mok ki, a Blackfeet chieftain, in 1802 for a trader working for the Hudson's Bay Company. On his return, Clark started preparing a massive map—it measures almost 3 x 5 feet (91 x 151 cm)—showing the West as a whole. He incorporated in it further information he had subsequently gained from trappers, traders, and Native Americans. It took ten years for a reduced and recompiled version of the original manuscript version to find its way into print. When it did, in 1814, its importance was recognized immediately. It became the foundation stone for all future mapping of the American West.

The Corps of Topographical Engineers

Following the success of the Lewis and Clark expedition, several others were launched. Two particularly notable ones in terms of their impact on American cartography were the 1809 exploration by Stephen Long (1784–1864) of the region between the Rocky Mountains and the Mississippi, and the search by Henry R. Schoolcraft (1793–1864) and James Allen (1806–46) for that river's source. Schoolcraft and Allen succeeded in their quest in 1832, when, although equipped only with a compass thanks to the meanness of the US Congress, Allen nevertheless accurately traced the expedition's route and drew the first map of the lake country in which the Mississippi has its origin.

The really great leap forward came in 1838, when Congress authorized the establishment of an expanded Army Corps of Topographical Engineers to replace the existing and inadequately staffed and resourced Topographical Bureau. From then on, surveying and mapping were both set on firm footings and the results were almost immediate and certainly impressive. Between 1842 and 1845, John Charles Freemont (1813–90), the so-called "Pathmarker of the West," then a lieutenant in the Corps, led three major surveying expeditions to the Rocky Mountains, Oregon, California, and Upper California. The Corps was also responsible for providing maps for the Mexican-American War of 1846 to 1848 and for organizing and carrying out the subsequent survey of the boundary between the two countries, which took from 1848 until 1853.

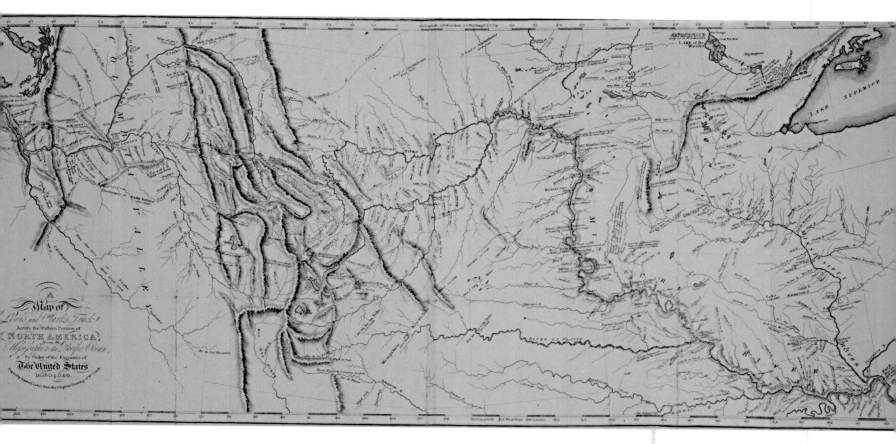

In 1853 the Corps was given what was to prove one of its greatest challenges. It was entrusted with the task of locating and then surveying the most feasible transcontinental railroad route from the Mississippi to the Pacific. Four surveying parties were sent out: one charged with preparing a northern survey, which concentrated on the 47th and 49th parallels, one on the 38th parallel, and two working along the 35th parallel from the west and the east respectively. These great surveys, combined with the survey of the Mexican border and those carried out for other railroad routes, marked a decisive moment in the mapping of the American West.

The lay of the land

Although the outbreak of the American Civil War (1861–5) put a temporary stop to the mapping of the continent, it soon resumed apace. Starting in 1867, two years after the war ended, and continuing into the 1870s, the federal government persuaded Congress to authorize the despatch of four more major surveys out West. Historians of cartography commonly refer to them by the names of their respective leaders: Clarence King (1842–1901), George Wheeler (1842–1905), Ferdinand Hayden (1829–87), and John Wesley Powell (1834–1902).

The first of these Great Surveys of the American West, as they are popularly known, was led by Clarence King. King and his expedition were given the task of surveying along the 40th parallel, between the Rocky Mountains and the Sierra Nevada. The survey was primarily focused on the geology of the region, although King was also charged with finding a possible alternate route for the mooted Pacific railroad. The area George Wheeler and his team were instructed to survey was much bigger, embracing California, Colorado, Montana, Idaho, Nebraska, New Mexico, Utah, and Wyoming. Wheeler made the production of a

74 The American West, Lewis and Clark, United States, 1814 Recording the travels of the Corps of Discovery led by US army captains Meriwether Lewis and William Clark in their exploration of America west of the Mississippi from 1804 to 1806, this map laid the foundations for all subsequent mapping of the American West. It was based on the daily route sketches and regional maps made by Clark during the course of the great expedition. Published in a reduced and recompiled version in 1814, it was the first map to survey the Missouri and Columbia River systems with any degree of accuracy, and the first to illustrate the northern part of the Rocky Mountains.

75 Railroad and Distance Map, Gaylord Watson, United States, 1871

Just two years after the historic completion of the Transcontinental Railroad in 1869, this wall map, produced by the New York City map, chart, and print publisher Gaylord Watson, reflected the way that the coming of the railroad was revolutionizing travel and communication in the United States. It advertised the fact that travelers could now journey from one side of the continent to the other in a matter of days rather than weeks, transforming the economy and outlook of the Pacific states. The company went on to produce more railroad maps, culminating in a *Railroad Atlas of the United States*. Though the company did not achieve the same commercial success as some of its rivals, its maps—particularly those of California and Nevada—were among the most detailed of their day.

Completion of the Transcontinental Railroad, USA, 1869
On May 10, 1869, an eastbound Central Pacific locomotive and a westbound Union Pacific locomotive met at Promontory Point, Utah, marking the completion of the United States' first transcontinental railroad.

detailed topographical map of all the territory west of the 100th meridian his personal goal. By the time the survey was completed, it had cost what was then the staggering sum of $691,444.45. During its course, 164 maps were produced, including 71 topographical maps drawn using a range of scales. These were collected together in a detailed *Topographical Atlas*. What made these maps particularly special was the eye-catching way in which relief was rendered through the use of hachures (lines indicating slope and height), plus their highlighting of routes—graded according to how easy it was to journey along them—through the Western terrain.

The atlas appeared in 1874. It was one of the eight volumes of Wheeler's massive final report, which, as a result of his illness, was not completely finalized until 1887. In addition to the atlas, the other volumes covered geology, zoology, palaeontology, botany, archaeology, astronomy, and barometric pressure, ending with a final geographical report. By that time, the army had lost a political and scientific battle for control of such surveys, which resulted in the Department of the Interior taking over responsibility for them from the War Department.

Civilian surveys

The Ferdinand Hayden survey of Colorado was one of the first to be sponsored by the Department of the Interior. The survey was completed in 1876, and the resulting *Atlas of Colorado* was published the following year. It contained four sheets of maps covering the entire state drawn at a scale of 12 miles to the inch, and 12 sheets—six geological and six topographical—at a scale of 4 miles to the inch. The atlas was notable for its economic map, which identified agricultural land, pasture, forest, and the locations of gold, silver, and coal deposits.

Another civilian survey, led by John Wesley Powell, began in 1871 and concluded in 1879. Powell's aim was to build on the survey he had already made of the Colorado River, extending it north of the Grand Canyon. Such was his success that, in 1882, Powell, then head of the recently founded United States Geological Survey, was given the authority to survey the entire country.

Coming of the railroads

Where the surveyors led, the "iron horse"—the name Native Americans gave to the locomotives of the new railroads—followed. Railroad expansion, once it got underway, was meteoric. In 1861, there were only 31,286 miles (50,350 km) of railroad track in operation; by 1871, this had increased to 60,643 miles (97,030 km), and, by the end of the century, to an astounding 186,809 miles (300,640 km), giving the United States the largest railroad system in the world.

American mapmakers, notably Chicago-based Rand McNally, were quick to capitalize on the railroad boom. The company was created in 1868, when William Rand (1828–1904), a Chicago printer, went into partnership with fellow printer Andrew McNally (d. 1905). The new company's first publication was the *Annual Report of the President and Directors of the Chicago, Rock Island, and Pacific Railway*. The company went on to become the biggest map publisher in the United States, building its success on the reports, tickets, maps, and guidebooks it published for the nation's burgeoning railroad companies.

Chicago was the natural base for such a concern, since it was a hub of the ever-expanding railroad network, with no fewer than 11 railroads converging on the city. By 1870, a train was leaving Chicago every 15 minutes. Rand McNally's first railroad map of the United States appeared in 1873, and many more were to follow. In 1876, the company produced its first large railroad wall plan: the New Railroad and County Map of the United States. Appearing just in time to mark the American centennial, it had taken a team of ten compilers and two engravers two years to create and had cost a total of $20,000 to produce.

MILITARY MAPPING

The mapping of the United States, and of the British Empire, owed an immeasurable debt to military surveyors and mapmakers. Such dedicated cartographers were continuing in a tradition that could trace its roots back to the end of the 15th century, when Charles VIII of France (1483–98) commissioned cartographer Jacques Signot to draw a map showing the king the best route for his artillery to take across the French Alps and into Italy.

From the late 16th century, military schools frequently offered mapmaking as part of their curriculum. At one such school, for instance, candidates for the army learned "enough drawing to design a fort and map a plan of campaign." In the main, however, commanders in the field seem to have relied on a mixture of graphic and oral information up until the middle of the 18th century.

The Battle of Culloden

For the generals of Louis XIV of France in the War of the Spanish Succession—as well as for the Duke of Marlborough (1650–1722) and Prince Eugene of Savoy (1663–1736), their chief adversaries—it seems to have been standard practice to obtain the best maps available of a given area and then to amplify their content by sending spies and reconnaissance troops forward, and gathering information about local conditions from the inhabitants.

This did not deter non-military mapmakers removed from the seat of war from attempting to map battles and campaigns. In Britain, for instance, 1705 saw the publication of a detailed True and Exact Map of the Seat of War in Brabant and Flanders with the Enemies Lines in their Just Dimensions, on which Marlborough's strategic moves and those of his French opponents could be charted.

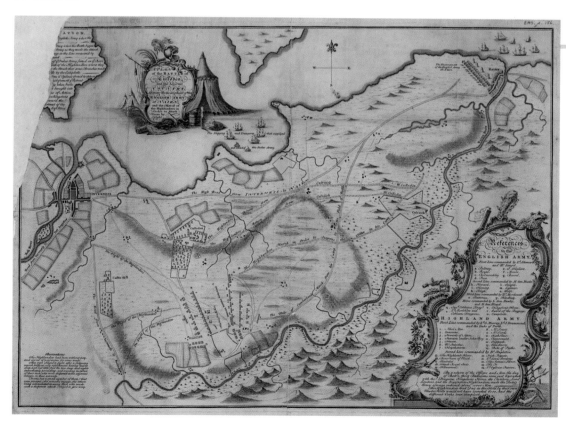

76 Plan of the Battle of Culloden, James Finlayson, 1746

James Finlayson served in Bonnie Prince Charlie's army as Ordnance Master of the Jacobite artillery, and his plan of the Battle of Culloden was considered by his contemporaries to be far more accurate than its various rivals. As well as mapping the decisive encounter itself, Finlayson showed the abortive night attack the Highlanders launched against the Duke of Cumberland's camp. The attack's failure is believed to have contributed greatly to the rout of the Jacobites the following day. Such mapping gave Finlayson's contemporaries, and the modern military historian, a vivid insight into the tactics and actualities of warfare.

Military men also frequently mapped specific land and sea battles after the event, the aim of such maps being to explain the context in which the particular battle had been fought as well as what had occurred during it. Thus, in Britain, the Battle of Falkirk, a major clash in the 1745 Jacobite rising against Hanoverian rule, was rapidly mapped following the Jacobite defeat at the Battle of Culloden in 1746. So, too, was Culloden itself. Probably the most celebrated map of it is the Plan of the Battle of Culloden, compiled by John Finlayson, the Jacobite artillery Ordnance Master in the battle (see p.139). A mathematical instrument maker by profession, Finlayson was also engineer and commissar to Prince Charles Edward Stuart (1720–88). Elements of the map— particularly in the decorative cartouche—are a clear indication of Finlayson's Jacobite sympathies.

A military revolution

Military mapmaking began to change at around the time Culloden was fought, becoming much more systematized and professional with the introduction of specialist draftsmen into the ranks of most armies, notably those of Britain, France, and Prussia.

In Prussia, the renowned soldier-king Frederick the Great (1712–86) instituted what he termed a *Plankammer*, a kind of traveling map office that accompanied his army on its campaigns. In his *Instructions for His Generals*, in which he drew on the lessons he had learned during the War of the Austrian Succession (1741–48), Frederick stressed the importance of consulting "the most detailed and exact maps that can be found," for "knowledge of the country is to a general what a rifle is to an infantryman and what the rules of arithmetic are to a geometrician." The rising awareness of what military mapping had to offer went hand in hand with other developments, such as the introduction of trained general staffs.

Mapping Waterloo

As armies grew larger and larger, so maps were used more and more effectively. Both Napoleon (1769–1821) and the Duke of Wellington (1769–1852) relied heavily on their respective cartographic staffs. In 1809, Napoleon set up an Imperial Corps of Surveys with a staff of 90, while, during the Peninsula War (1808–13), Wellington took a mobile printing press with him to Spain, so that new maps could be compiled and printed quickly. The French emperor himself never embarked on a campaign without taking a mobile map collection with him in a special cart. Contemporary accounts reveal that Napoleon also spent a great deal of time minutely examining maps specially drawn up for him by the eminent mapmaker General Louis Bacler d'Albe (1761–1824), his cartographic expert, together with other maps prepared by the topographic office of the Grande Armée. He used these maps to plot his dispositions and chart those of the enemy.

For his part, Wellington made extremely effective use of the maps he had requested his cartographic staff to prepare for him immediately prior to the clash with Napoleon at Waterloo in 1815, the emperor's final defeat. Knowing that the French would inevitably march north against the British and their allies, who were concentrated around Brussels, before they could be reinforced, Wellington had maps drawn of four or five possible battlefields, so that he could plan the disposition of his forces in advance. The Sketch of the Ground Upon which was Fought the Battle of Waterloo, as it was somewhat cumbersomely titled after the event, proved to be the crucial one. The original bears pencilled marks, thought possibly to have been made by Wellington himself, indicating how he intended to position his troops for the great battle. After the battle, an officer serving in the Royal Engineers used a copy of the sketch to record the events of the actual battle. His careful color-coding showed the French forces taking the offensive, while the main Allied army waited in a carefully chosen defensive position just over the brow of a ridge.

The Kings' Cake being Cut at the Congress of Vienna, 1815
A French political cartoon shows Europe's rulers greedily redrawing the map of Europe at the Congress of Vienna, before Napoleon's escape from Elba in February 1815.

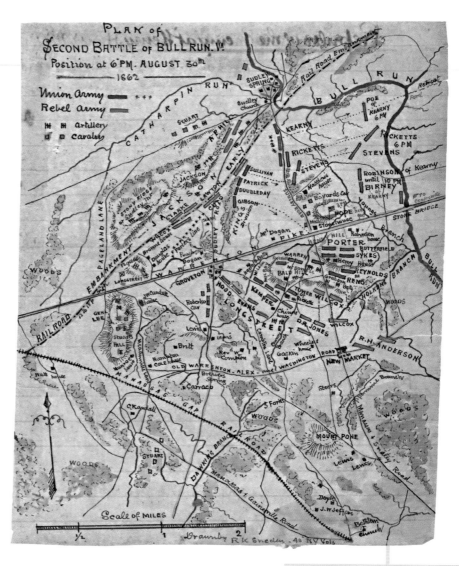

The American Civil War

In the 19th century, warfare can be argued to have encouraged the growth of public interest in maps, and the American Civil War was no exception. Between 1861 and 1865, the year the South was forced to capitulate, Union and Confederate topographical engineers surveyed thousands of miles of the South, including many areas that had never been charted. Albert Campbell, the head of General Robert E. Lee's Topographical Department, and Jedediah Hotchkiss, a brilliant amateur who became chief cartographer of the Army of Northern Virginia, were probably the best-known Confederate cartographers, while, in the North, William E. Merrill, chief topographical officer with the Army of the Cumberland, operated what is often held to have been the most sophisticated field mapping unit of either side in the Civil War.

At the same time, map publishers and newspapers on both sides were providing the general public with specially drawn maps of battles, campaigns, and the overall theaters of war. Between April 1861 and April 1865, for instance, it has been calculated that daily newspapers in the North printed no fewer than 2,045 maps relating to the war. The Confederacy produced far fewer such maps, probably because of lack of the skilled labor and the printing materials required to make and produce them.

Such maps were produced with one aim in mind: to enable people at home to follow the movements and events of a war in which it was more than likely that they had relatives and friends actively involved. Sometimes, the desire to keep the public informed outweighed military commonsense. In December 1861, for instance, the front page of the *New York Times* featured a detailed map of the Union defenses of Washington and the forces holding them. General George B. McClellan (1826–85), the commander of the Army of the Potomac, demanded the newspaper's immediate suppression as punishment for aiding the Confederacy.

THEMATIC MAPS

The Age of Empire saw a growth in what can best be termed thematic mapping. Such maps concerned themselves with the presentation of subjects such as the new science of demographics and other kinds of information that lend themselves readily to depiction in map form. Maps have been thematic, in their inclusion or exclusion of different elements, since the very earliest times, but the growth of scientifically inspired thematic cartography changed the way we present and understand information for ever.

On the chart of the Gulf Stream drawn by Benjamin Franklin (1706–90) in 1769, for instance, arrows are used to indicate the direction of flow of the current, while the American oceanographer Matthew F. Maury (1806–72) won the gratitude of generations of sailors for the

77 Second Battle of Bull Run, Robert Knox Sneden, United States, 1862
Drawn on the battlefield, this map shows the Union and Confederate positions in the evening of August 30, 1862, during the Second Battle of Bull Run, which, like the first, ended with a heavy Union defeat. The color coding indicates the location of the Northern and Southern forces. Sneden enlisted as a private in the Army of the Potomac before becoming a Union mapmaker and, after being captured in a Confederate raid, a prisoner-of-war in Andersonville, the South's most notorious prison camp. His fascinating maps and drawings were rediscovered only in the 1990s.

78

Delineation of the Strata of England and Wales with Part of Scotland, William Smith, England, 1815

William Smith began work on the first geological survey of England and Wales in 1805. His method involved the laborious recording and comparing of the fossil content of one layer of rock with that of another. Different rock formations were distinguished by use of color. The tonal variations indicate the different levels of rock formations: the paler the tone, the deeper the rock. Despite all Smith's efforts and the support of the Royal Society, the survey was not a commercial success when it was published in 1815. It took some time for it to be recognized as a milestone of mapmaking.

so-called track charts he devised together with his global map of the winds and ocean currents of the world. The latter showed the prevailing trade winds of the oceans, the seasonal migration of the doldrums, and the best routes for finding the most favorable winds.

Other important examples of early thematic mapping include the 90 thematic maps, covering meteorology, climatography, hydrology, hydrography, geology, earth magnetism, botanical and zoological geography, anthropography, and ethnography, that the German geographer Heinrich Berghaus (1797–1884) was inspired by Alexander von Humboldt (1769–1859) to create for his *Physikalischer Atlas*, the first volume of which appeared in 1845 and the second three years later. This early thematic atlas is generally regarded as one of the most extensive and detailed ever to be produced. For his part, Humboldt, lauded by Charles Darwin (1809–82) as "the greatest scientific traveller who ever lived" and one of the undisputed founders of modern geography, drew several thematic maps during the course of his expeditions, including what is termed an "altitude profile" of equatorial plants.

Thematic cartography also played a major part in the emergence of the new science of demography. Pioneers of demographics included social commentators such as Francis Amasa Walker, the head of the US Bureau of Statistics who introduced maps for the first time into the *Report of the 1870 Census* he superintended. He went on to produce the *Statistical Atlas of the United States*.

Geological maps

Geological mapping is an example of the development of thematic mapping. William Smith (1769–1839) employed what is technically termed the principal of fossil selection—that is, the comparing of the fossil content of one layer of rock with that of another—to create the first geological record of his homeland, A Delineation of the Strata of England and Wales with Part of Scotland. It was a massive project, and the resulting map, drawn at a scale of 5 miles to the inch, was truly imposing. The 15 sheets of which it is composed measure 6 x 9 feet (182 x 274 cm) when laid side by side.

Smith's fascination with fossils began when he was a small boy on his uncle's Oxfordshire farm. Later he became a largely self-taught surveyor, traveling all over England and then, in 1796, starting to draw up local geological maps from his base in Bath. Just under a decade later, he set out to map the rock strata of the whole country, using fossil observations as his starting point. Progress was slow, and the money Smith needed to finance the map's preparation and eventual publication proved hard to come by, even though he had won the support of Sir Joseph Banks, the president of the Royal Society, and John Cary, one of the most important British map-publishers of the day. It was Cary who provided Smith with a topographical base map of the country, an essential prerequisite to the creation of the geological survey.

Production of the map eventually began in 1812, and three years later it was published in a limited edition of 400 copies. Unfortunately for Smith, it failed to sell. He had to dispose of his fossil collection to the British Museum to raise funds, and was even briefly sent to a debtor's prison. In 1819, he resumed mapmaking, issuing a reduced version of his original map and producing a series of geological maps of individual English counties. It was not until late in Smith's life that his contribution to geology came to be appreciated. In 1831, the Geological Society of London awarded him the first ever Woolaston Medal. That was—and is—the highest honor the Society can bestow for achievement in geology. Following in his footsteps, official mapping got under way in the 1830s and, by 1890, the geological mapping of England and Wales was complete.

New ways of presenting statistics

In 1861 French statistician Charles Minard (1781–1870) drew a map of Napoleon's march to Moscow in 1812 and subsequent retreat from the Russian capital. In it, what is basically a geographical map is overlaid with a huge amount of statistical information, much of which is conveyed in graphics. The map was one of a series of *cartes figuratives* (figurative maps) that Minard had begun to produce from the mid-1840s onward. Other celebrated examples of Minard's mapping include the flow map he devised in 1862 to chart patterns of European emigration, a map detailing in terms of volume how much wine France was exporting, and a map showing cotton imports to Europe.

Minard's requirements often led him to ignore or distort geographic reality—as in his 1866 map showing the trade in raw cotton before, during, and after the American Civil War—for which he was much criticized at the time. Minard summed up his approach as being: "less to express statistical results, better done by numbers, than to convey promptly to the eye the relation not given quickly by numbers requiring mental calculation."

Mapping social conditions

Many thematic maps were a cartographic response to the process of industrialization, the key socio-economic driver of the entire period. In the main, such maps were a reflection of the growing social problems that accompanied urban growth and social stratification. With the dramatic shift in population from the country to cities and towns, some form of civic planning eventually became essential, even though the development was at first resisted by the *laissez-faire* capitalists of the day.

In Britain, for instance, an uncontrolled, unplanned, and unmapped expansion resulted in burgeoning, prosperous Liverpool having some of the worst living conditions in the country. At the same time, however, there was the Victorian desire to erect great municipal buildings. These reflected a sense of civic purpose, even though the development did little to improve the lot of the ordinary people. It was not until after the turn of the century, with the passage of the 1909 Housing and Town Planning Act, that urban planners were given the tools to tackle the problem. The Act allowed local authorities to prepare development schemes for land close to towns.

One area that particularly concerned 19th-century philanthropists was the impact of social conditions on public health. At the same time, maps began to be employed as a means of mapping outbreaks of disease, and as a means of combating them. In 1798, Valentine Seaman (1770–1817) produced the first maps to show the incidence of diseases when he used dots and circles to indicate individual occurrences of yellow fever in the waterfront areas of New York.

In Britain in 1832, the Poor Law Commission published its *Report on the Sanitary Condition of the Labouring Population*, in which maps of housing types and of the incidence of disease in Leeds, Yorkshire, and Bethnal Green, London, were featured. In the same year, American mapmaker Henry Schenck Tanner (1786–1858) produced a detailed study of the worldwide cholera epidemic, which had reached the United States from India, where it had originated in 1817. Tanner depicted the spread of the disease in a world map, supporting it with more detailed maps of the United States and of New York, showing where the disease had struck to establish that there was a geographic connection underlying its dissemination.

Cholera was one of the 19th century's chief killers. The problem confronting the doctors who were attempting to treat it was that they had no real idea of how it was contracted. Then, in 1855, John Snow (1813–58), an eminent London doctor, published *On the Mode of Communication of Cholera*. In it, he included two maps, showing how the incidence of the disease varied depending on which water company was supplying the area in London's Soho, where an epidemic had started the previous year, and how the known cholera cases appeared to be clustered around specific public water pumps, notably the one in Broad Street. The task had involved Snow in surveying the area household by household to establish where each obtained its drinking water, and then correlating this information with the addresses of all the known cholera victims. The results were conclusive: contaminated water from the pump was to blame. The supply to the affected pump was provided by the Southwark & Vauxhall Water Company, which drew its water from the River Thames not far from where the main sewers discharged their effluent into the filthy river.

The power of demographics

Social reform went hand in hand with health reform. By the second half of the 19th century, London, with a population approaching four million, was the world's largest and richest city, yet many of its people, as noted by Karl Marx (1818–83), lived in appalling conditions and abject poverty. The millionaire shipping owner and social reformer Charles Booth (1840–1916) commissioned a series of maps to chart both poverty and wealth.

Disbelieving the claim that a quarter of London's population lived in poverty, Booth decided to carry out a comprehensive survey of the city, in which he combined statistical methods with direct observations. The result was a monumental analysis of late Victorian social conditions, which Booth drew on freely when he came to write *Life and Labour of the People* in 1889. This dealt with the East End. A second instalment, *Labour and Life of the People of London*, which appeared in 1891, covered the remainder of the city.

Maps featured heavily in Booth's work. What he did was to color-code areas according to their social classification, ranging from black for the lowest-class areas and light blue for the very poor ones to red for the middle class and yellow for the very wealthy (see pp. 146–7). What came as a surprise to both Booth and his readers was that, taken as a whole, a staggering 35 percent of Londoners could be said to be living in want and that, paradoxically, the people he characterized as being of "the lowest class" and as "vicious semi-criminals" frequently lived in close proximity to the wealthy.

Booth's great work became the prototype that all similar sociological surveys were to follow in the future. His enlightened conclusion was that, despite the moralistic tone of some of his language, the only way of reducing crime—a major Victorian concern—was by reducing poverty. He also concluded that poverty was inevitably a consequence of unemployment and was age-related. This led him to become a champion of state pensions for the elderly, and he was to live to see these come into effect.

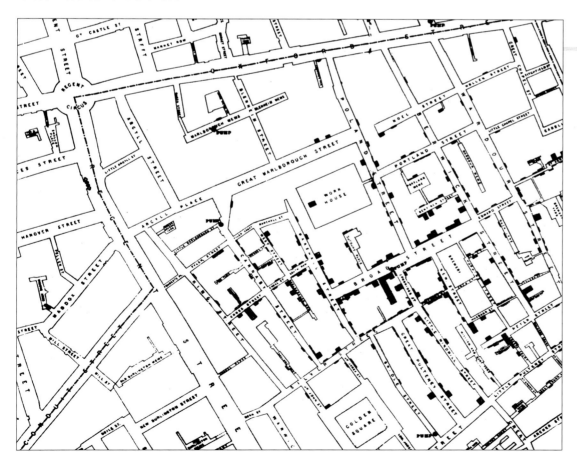

79

On the Mode of Communication of Cholera, John Snow, England, 1855

Leading London physician John Snow used two maps of Soho in his classic treatise on the reasons for the spread of cholera. The maps showed that the distribution of cholera cases was directly related to the positioning of the public water pumps in the area and the sources of their water supply. Cases of the disease are shown by black bars on the map. Snow's findings proved conclusively his theory that the distribution of polluted drinking water was the key reason for the incidence and spread of the disease.

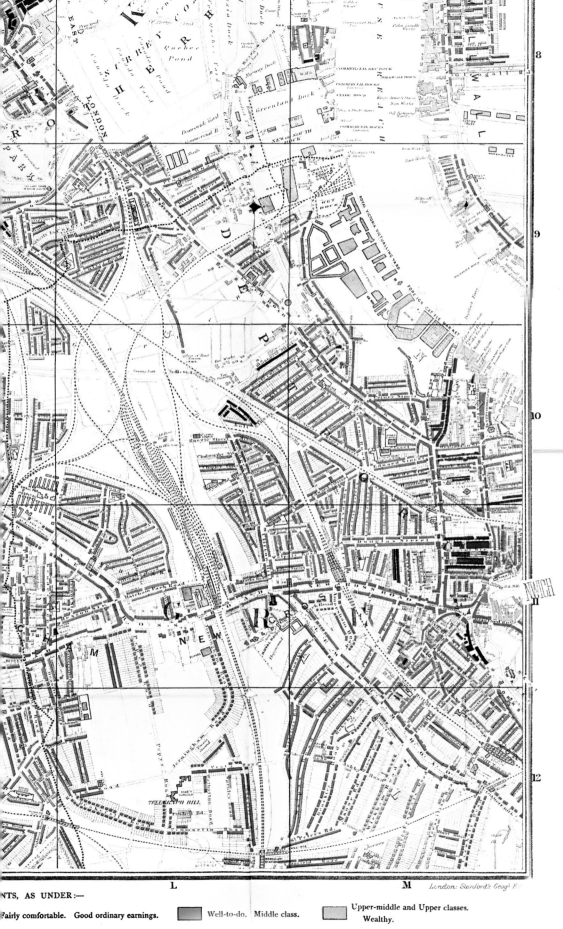

80

Descriptive Map of
London Poverty,
Charles Booth,
England, 1889

Victorian social reformer and shipping
tycoon Charles Booth devised this map to
accompany his groundbreaking survey of
Labour and Life of the People of London,
which highlighted the prevalence of
poverty among the city's teeming
millions. The map used color to define
what Booth termed "the general
condition" of the city's inhabitants area
by area and street by street. There were
seven possible classifications, ranging
from "Vicious, semi-criminal" to "Upper-
middle and Upper Classes. Wealthy."
Based on the careful collection of masses
of statistics, the map made it abundantly
clear that poverty was a major social
issue, with 35 percent of Londoners
living below what Booth defined as the
poverty line.

NTS, AS UNDER:—

Fairly comfortable. Good ordinary earnings.   Well-to-do. Middle class.   Upper-middle and Upper classes.
Wealthy.

represented by the respective colours.

IT'S A LONG WAY

TO

.ROME

191ᵉ JOUR
1-4-44

320 KM

180 KM

DÉPART
6-9-43

RECORD DE LENTEUR

KM

la plus grande vitesse atteinte par l'escargot est de 0ᵐ80 à la minute.

THE MODERN WORLD: New Visions

During the 20th century, mapmaking went through a seismic change. As they became less costly and easier to reproduce, maps became part and parcel of the fabric of everyday life, with a hungry and ever expanding audience for them emerging across the globe. In response, more maps than ever before in history were created to satisfy the requirements of the consumer marketplace and also to service new needs.

The advent of the computer age also encouraged the spawning of many new maps. Some would say that computerization, along with the development of map-specific hardware and software packages, were the catalysts that sparked off a new cartographic revolution. The arrival of such technology meant that maps could be generated faster and in greater quantity, even when complex collections of data were involved. For cartographers, the digitization of such information made the entire mapmaking process much easier. Instead of laboriously having to collect, collate, and transcribe individual elements, it now takes a matter of minutes to locate and download the necessary information from digital databases. Others would argue that computerization has made maps less artistic and thus less appealing.

The first computer-enhanced maps to be created were weather maps, which were generated by ENIAC (Electrical Numerical Integrator Computer)—the grandfather of all modern computers—as early as 1950. This was an enormously important development, since an improved ability to map the weather meant that it became more practical to predict the onset of violent storms, such as hurricanes, tornadoes, and cyclones. Just over 20 years later, the first prototype electronic atlas appeared, followed by the first CD-Rom map of the world in 1989.

The grandfather of computers, ENIAC, 1946 Inventors J. Presper Eckert and J.W. Mauchly work on the Electrical Numerical Integrator Computer (ENIAC) at the Moore School of Electrical Engineering at the University of Pennsylvania.

81 Italian Campaign Propaganda Map, Germany, 1943

Produced by the Nazis for publication in Occupied France and for airborne distribution in Italy, where Free French troops were fighting as part of the Allied invasion forces, this map highlights the slow progress of the Allies from their original landing point in the Bay of Naples north up the Italian peninsula, pointing out that it was still a long way to Rome. Having landed in Italy in the autumn of 1943, it took the Allies until June 5, 1944—the day before D-Day—to capture Monte Cassino, break through the German Gustav Line, and eventually enter the Italian capital. Maps such as this were often produced during World War II as propaganda tools to point out to the troops on which they were airdropped the "true" strategic and tactical position. In May 1940, for instance, the Germans produced maps showing the British and French forces at Dunkirk that they were encircled in a bid to get them to lay down their arms.

LEISURE AND TRAVEL

In the Western world, growing prosperity together with the advent of affordable, reliable family automobiles and, from the second half of the 20th century, increasingly inexpensive air travel, meant that more and more ordinary folk were now able to venture farther, faster, more often, and more conveniently than ever before. The immediate

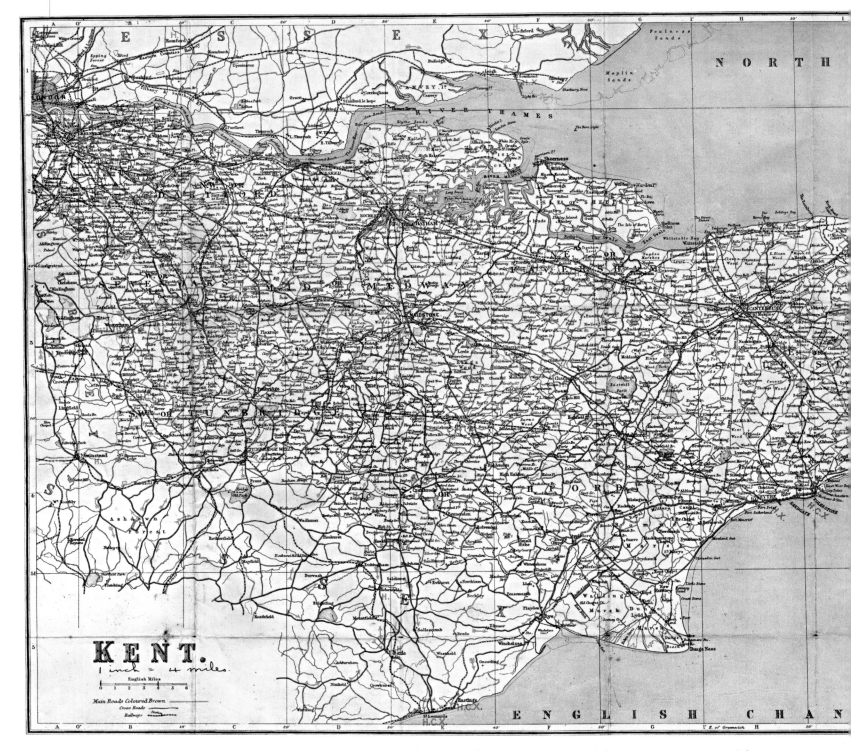

KENT.

1 inch = 4 miles.

English Miles

Main Roads Coloured Brown
Cross Roads
Railways

consequence was the rapid growth of popular tourism and the consequent demand for maps specifically tailored to meet leisure travel needs.

When cycling took off in Britain in the late 1880s, the supposedly staid Victorians were quick to jump on their bikes and head for the countryside. By 1893, some 500,000 dedicated British cyclists were taking advantage of the freedom the bicycle gave them; within a decade, the number of cycling enthusiasts had more than doubled to well over a million. The pioneering maps produced by cartographers such as the Scottish map publisher John Bartholomew gave these cyclists the means to explore the land to the full.

Other influential cycling maps came from George Philip & Sons, founded in Liverpool in 1834 by George Philip (1800–82) and one of the major British map producers of the late 19th and the 20th centuries. Among its products, the company originated *Philip's Waterproof Maps for Cyclists*

82
Kent, *Cyclist's Maps of the Counties of England*, George Philip & Son, England, 1905

Like its Scottish rival, John Bartholomew, the London-based map publisher George Philip & Son was quick to capitalize on the British cycling boom at the end of the 19th century. This map of Kent—the "Garden of England"—was one of a series of cycling maps produced by the company in the early 20th century. The cartography was based on the Ordnance Survey with annotation added to make the maps even more useful. The "C" stands for Consul of Cyclists Touring Club, "H" for recommended hotels, and "X" for where cycles could be repaired. The arrows relate to hills.

in 1899. These, it claimed, were "impervious to water" and "readily cleaned with a sponge." They also "took up less room" than their rivals and came complete with a guarantee of "greater satisfaction."

Nor was the cycling boom confined to Britain. In the USA, the quaintly named League of American Wheelmen—now the League of American Bicyclists—encouraged its affiliated clubs to create local logs of cycle routes, from which elaborate route maps, such as George W. Blum's *Map of California Roads for Cyclers*, were eventually to be compiled. Blum went on to produce a *Cycler's Guide and Road Book of California* in 1896. It contained three maps—one of California as a whole, a second of Golden Gate Park in San Francisco, and a third showing the route from Chico to San Diego—plus an itinerary of cycle trips throughout the state.

Back across the Atlantic in Italy, a group of cycling enthusiasts founded the Touring Club Italiano in 1894. Widening its scope to promote tourism generally, the club published its first set of maps, the *Carta Turistica d'Italia*, in 1909. These were among the first of their kind to be created specifically with the needs of tourists in mind and set the standard that all future maps had to match. Harking back to pre-Renaissance traditions, such maps frequently draw important buildings and tourist sites in pictorial form (see p.152).

Oil company maps

Where cyclists led, motorists were quick to follow, notably in the USA, where the major map publishers were quick to capitalize on motoring's commercial potential. The first road atlas of the country, *A Survey of the Roads of the United States of America*, had been published as far back as 1789. Now, with the coming of the automobile, a plethora of new road maps followed, many of them being given away for nothing by touring organizations, automobile clubs, oil companies, and the nationwide networks of gas stations.

The General Drafting Company issued its first road map, of Vermont, in 1912 for the newly established American Automobile Association, while in 1917 Rand McNally began to publish what would develop into the first national series of road maps. The company called the series *Auto Trails Maps: A Guide to the Blazed Trails*. Rand McNally's interest in road maps dated back to 1907, when Andrew McNally II, the grandson of the co-founder of the company, decided to drive his new bride from their hometown in Chicago to spend their honeymoon in Milwaukee. Combining business with pleasure, he strapped a camera to the front of his car and stopped at every junction and turning to snap a photograph. Back in Chicago, he compiled the photographs into a booklet, with a little arrow added to each photograph to indicate which road to take. McNally titled the booklet a *Photo-Auto Guide*.

By the mid-1960s, over 200 million free oil company maps were being published annually by Rand McNally, the H.M. Gousha Company, and the General Drafting Company. By the 1970s, when the practice started to come to an end primarily as a consequence of the impact of the rise

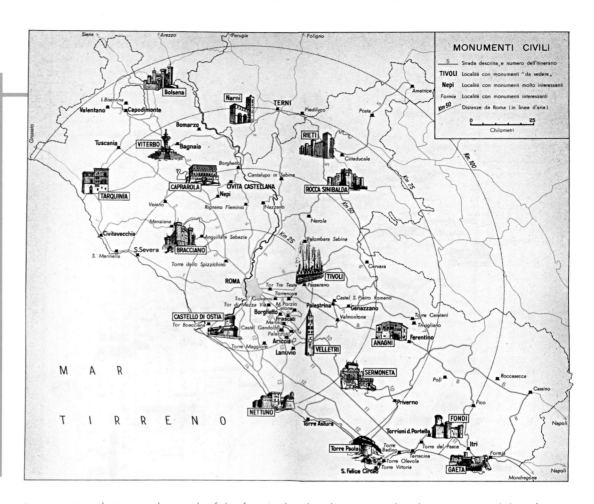

83 Tourist Map of Lazio, Touring Club Italiano, Italy, 1953 A handy traveling reference to the tourist sites of the Lazio region, this map was one of many published by the Touring Club Italiano from the 1920s onward as guides to locating the best of the country's architectural, artistic, and cultural heritage. Like many tourist maps, the plan presents key sites in pictorial form to aid navigation. Founded in the 1890s by a band of cycling enthusiasts, the club grew in importance and prestige with the rise of motoring and the consequent growth of tourism. The club issued its first maps in 1909; in 1922, it published the first guidebook to Italy designed for use by overseas visitors.

in gas prices that were the result of the first Arab oil embargoes, it has been estimated that, from the 1920s onward, more than 10 billion free road maps had been produced for and given away to the ever-increasing number of American motorists, helping road travel for generations of motorists.

In Europe, such maps started to appear in around 1930, though, with the exception of Germany, the oil companies produced relatively few of them before World War II. They were at their peak from roughly 1952, when Esso established its Esso Touring Service, to 1972. Together with Esso, Shell dominated the European gas map market, followed by BP and Aral.

The oil companies did not monopolize this field, however. Other concerns, such as the Michelin Tire Company and motoring organizations, like the AAA in the USA and the AA in the UK, also produced motoring maps, as did clubs such as the Touring Club Italiano in Italy.

NEW VIEWS

Advances in technology deeply influenced how map data were gathered and ultimately the ways in which maps were produced. The coming of photogrammetry—the use of photographs, especially aerial ones, to make accurate measurements—meant that it was no longer essential to send large parties of surveyors and mapmakers out into the field to prepare basic topographic maps. Cameras could make a permanent record of the necessary detail far more quickly, efficiently, inexpensively, and accurately. They also made it possible to scrutinize the landscape from varying heights and different angles. However, the basic geodetic framework still had to be surveyed instrumentally on land.

Aerial photography

Aerial photography started as far back as 1858 when the pioneer French photographer and balloonist Felix Nadar (1820–1910) photographed Paris from a balloon drifting high above the city. His goal was to produce land surveys from the aerial photographs he took. During the American Civil War, the legendary cavalryman George Armstrong Custer (1839–76) created some of the world's first maps sketched from the air when, as a young officer fresh from West Point, he was attached to the Union Army's experimental balloon corps. In 1909, Wilbur Wright (1876–1912) took the first aerial photographs from an airplane while flying over Centocelli, Italy.

Then came the impact of world war. By 1918, French military reconnaissance aircraft were taking 10,000 aerial photographs of the Western Front a day alone, while units of Britain's Royal Flying Corps were experimenting with the use of aerial photographs as the basis for producing maps of Turkish-held towns in Palestine. In the UK itself, though, aerial photography was not adopted for civil mapping until World War II was nearing its end.

Things moved much faster in the USA. In 1921, the US Geological Survey used 274 aerial photographs specially taken by a pilot in the Army Air Service to map a 225 square mile (582.7 sq km) area near Kalamazoo, Michigan: the so-called Schoolcraft Quadrangle. As far as the USA was concerned, this was the first map in cartographic history to rely solely on aerial photography for its compilation. Its publication marked the start of a revolutionary new era in mapmaking.

Following this, the USA also pioneered the development of so-called orthophotomaps—aerial photographs with their backgrounds rectified to remove image displacements and then enlarged to a constant scale. The advantage of such maps is that, unlike aerial photographs, accurate measurements can be made on them, while cartographic elements, such as a grid, contours, spot heights, and place names, can be added.

Under the sea

Other technological advances—many of which were sparked off by the two world wars and their aftermaths—made it possible for cartographers to map phenomena that it had hitherto had been impractical or impossible to record. Without the development of improved forms of sonar, the underwater equivalent of radar, during World War II, for instance, it would have been impossible for the American oceanographer Marie Tharp (b. 1920) and her collaborator earth scientist Bruce Heezen (1924–77) to obtain the geophysical data they needed in order to produce their Physiographic Map of the North Atlantic, which they published in 1957. Tharp prepared the map, while Heezen collected the ocean floor data from profiles of the sea bottom made by echo sounder.

The result was the first map of the sea floor that enabled viewers to visualize the bottom of the ocean. Together with their 1977 World Ocean Floor Panorama, which illustrated how earthquakes follow the Earth's shifting tectonic plates, it helped to provide a logical explanation for such phenomena as the formation of mid-ocean ridges and trenches and the existence of the "ring of fire" around the Pacific.

84 Pennsylvania Automobile Road Map, Gulf Refining Company, USA, 1916
The Gulf Refining Company, formed in 1901, was one of the earliest oil companies to distribute free road maps, by mail and from its gas stations, from about 1913. This map of Pennsylvania shows the "best" routes in red. The numbers on the roads correspond to driving instructions published by the Automobile Blue Book Publishing Company, as the first route signs did not appear until after 1917, and route numbers did not start to be implemented until 1925.

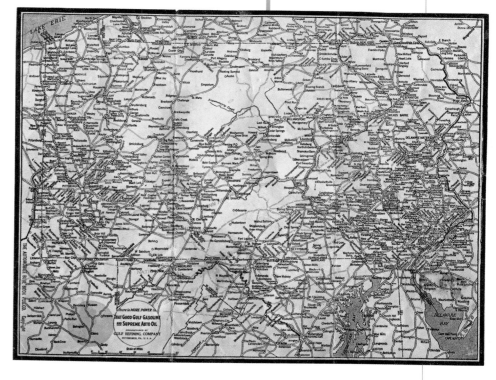

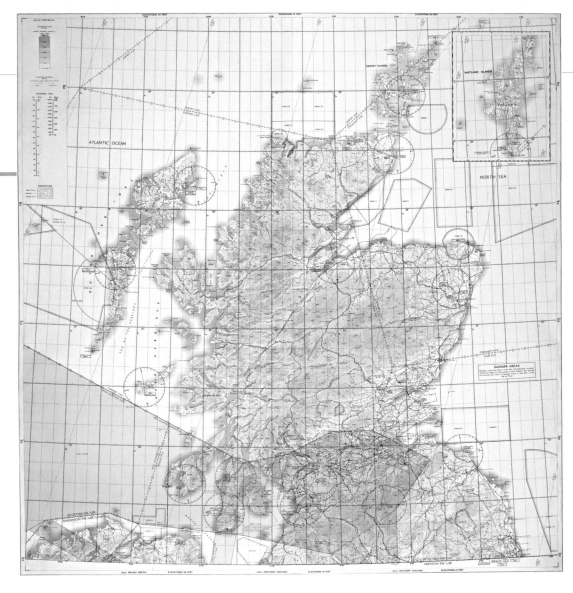

85 Aeronautical Chart of Scotland, Civil Aviation Authority, England, 1972

When airliners first took to the skies in the 1920s, aerial navigation was in its infancy. Night flying was practically unknown and, when it came to finding the way by day, it was common practice for pilots to drop down and look for a railroad line or landmark to check their bearings. By the time civil aviation really took off in the decades after World War II, aeronautical mapping had been revolutionized. Detailed charts were developed to aid in flight planning and navigation. Their development was a vital key to maintaining air safety.

Similarly, it would have been impossible to gather the information that Tharp, Alvaro Espinosa, and Wilbur Rinehart needed to compile their groundbreaking World Earthquake map of 1981 without the assistance of the new global network of seismic monitoring stations. Based on records of an amazing 56,000 earthquakes over a 20-year period from 1960 onward, the result was a vivid global depiction of the great earthquake zone that snakes its way around our planet.

Such mapping has greatly aided the ability to predict tsunamis and their likely consequences. These terrifying waves can be generated as a result of earthquakes, volcanic eruptions, underwater landslips, or the impact of giant meteorites, the result being a major displacement of the tectonic plates in the affected area. The waves themselves are truly awesome—in the case of the tsunami that struck Indonesia and other parts of southern Asia in December 2005, it has been estimated that they rose to a height of 29 feet (9 m) and traveled at speeds of 1,000 miles per hour (800 km/h). The more cartography reveals about the nature of the ocean floor, the more practical it is becoming to develop faster and more reliable ways of forecasting such terrible events.

THE WORLD WARS

For cartography, the two world wars and subsequent conflicts were powerful catalysts. Newspapers, magazines, and, later in the century, television turned to maps to help to illustrate the war news, with maps detailing the various theaters of operations. Such mapping encouraged people at home to track the progress of their troops, and could have a vital propagandist role. For their part, soldiers, sailors, and aviators benefited from new and better mapping.

As the century progressed, military commanders became increasingly reliant on providing soldiers, sailors, and airmen with accurate maps and the training to use them. The French had learned this lesson as early as the Franco-Prussian War of 1870–71, when Napoleon III's optimistically named Army of the Rhine set out to meet its foes plentifully equipped with detailed maps of Germany, but few, if any, of the French border regions, where the decisive part of the action was to take place, thus contributing to the defeat of France.

During World War I, the advent of trench warfare meant that mapmakers were called upon to create thousands upon thousands of detailed large-scale maps, without which it would have been impossible to construct effective defenses or to plan attacks on them. Infantrymen, artillerymen— and, of course, the generals and their staffs—all became map-conscious and map-dependent. In 1914, the British Expeditionary Force (BEF) in France started off with a mapping staff of a solitary officer aided by a single clerk; by 1918, the BEF possessed a 5,000-strong surveying department, and more than 35 million maps of its sector of the Western Front had been produced.

Impressive though this figure sounds, the mapping output of World War II dwarfed it. Between 1939 and 1945, it is estimated that, on the Allied side alone, more than a billion different military maps were created. The American Army Map Service was responsible for more than 500 million of them, and Britain's Ordnance Survey for a further 300 million. One major contributing factor in this dramatic escalation was the emergence of air power as a potent weapon in its own right. British Prime Minister Stanley Baldwin might well have commented pessimistically in 1935 that "the bomber will always get through," but the realities of actual aerial warfare—especially the demands of long-range night flying—meant that specially detailed maps were vital necessities if the bombers were successfully to locate and hit their targets.

Aeronautical charts

The German Luftwaffe led the way in aeronautical mapping, perhaps not surprisingly, since the first air navigation maps in history had been devised by Lieutenant Colonel Hermann Moedebeck (1857–1910), a Prussian artillery officer and one-time commander of an army balloon detachment, who produced a set of them for Count von Zeppelin's infant airship company in 1909. As well as developing an early form of electronic guidance for its bomber crews during the Blitz on Britain in the autumn and winter of 1940 and the spring of 1941, the Luftwaffe provided them with Ordnance Survey-derived maps of their targets, specially printed on plastic-coated fabric to make them easy to unfold, fold, and manipulate in cockpit conditions.

The cities and towns that formed the Luftwaffe's targets were highlighted in vivid yellow with red surrounds, the intention being to make them luminous in total darkness. Where practical, the Germans also enhanced their maps with information obtained from detailed photoreconnaissance. The British RAF and the US Army Air Force were quick to follow and improve on the German example for their own bombing campaigns.

Mapping D-Day

Each arm of the fighting forces had its specific mapping requirements. These became even more demanding and complicated when all three branches were working together in the kind of combined operation of which D-Day, the Allied invasion of France in June 1944, stands as the prime example.

With around 6,000 vessels, ranging from mighty battleships to small landing craft, ferrying 150,000 soldiers across the choppy waters of the English Channel in the dark to face the cream of the German Wehrmacht marshaled behind its much vaunted Atlantic Wall, British premier Sir Winston Churchill's description of the enterprise as "undoubtedly the most complicated and difficult"

American troops landing in Normandy, D-Day, June 6, 1944
American troops in landing craft go ashore on 3.5-mile (5.6-km) Omaha beach. Within a few hours of landing, the Americans had suffered more than 2,400 casualties on Omaha.

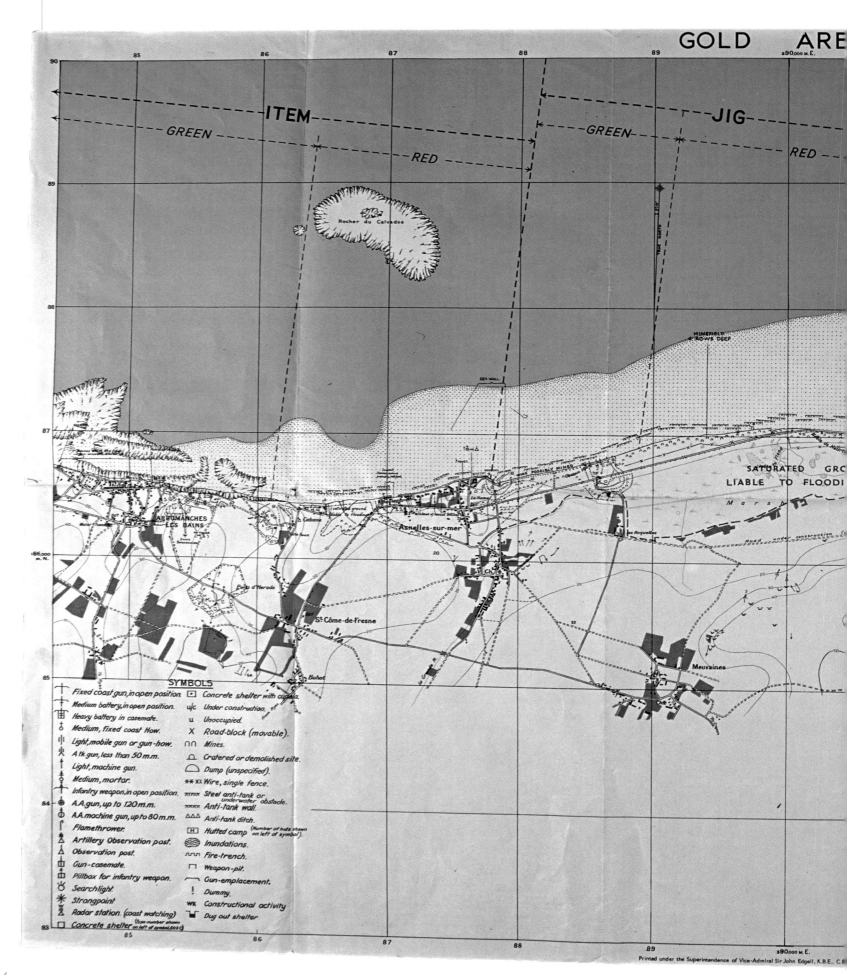

ITEM

GREEN

RED

JIG

GREEN

RED

Rocher du Calvados

MINEFIELD
4 ROWS DEEP

SEA WALL

SATURATED GRO
LIABLE TO FLOODI

Marsh

ARROMANCHES
LES BAINS

Cabane

Asnelles-sur-mer

Les Roquettes

Road under construction

Puits d'Herode

Ch

St Côme-de-Fresne

Meuvaines

Buhot

SYMBOLS

Fixed coast gun, in open position.		Concrete shelter with cupola.	
Medium battery, in open position.	u/c	Under construction.	
Heavy battery in casemate.	u	Unoccupied.	
Medium, fixed coast How.	X	Road-block (movable).	
Light, mobile gun or gun-how.		Mines.	
A tk.gun, less than 50 m.m.		Cratered or demolished site.	
Light, machine gun.		Dump (unspecified).	
Medium, mortar.		Wire, single fence.	
Infantry weapon, in open position.		Steel anti-tank or underwater obstacle.	
A.A.gun, up to 120 m.m.		Anti-tank wall.	
A.A.machine gun, up to 80 m.m.		Anti-tank ditch.	
Flamethrower.		Hutted camp (Number of huts shown on left of symbol).	
Artillery Observation post.		Inundations.	
Observation post.		Fire-trench.	
Gun-casemate.		Weapon-pit.	
Pillbox for infantry weapon.		Gun-emplacement.	
Searchlight.		Dummy.	
Strongpoint	wk	Constructional activity	
Radar station. (coast watching)		Dug out shelter	
Concrete shelter (type number shown on left of symbol,503)			

KING

GREEN

GREEN

RED

GREEN

O.N.1 APPENDIX VII.
ANNEXE A
INFORMATION UP TO 6th APRIL 1944

LOVE

N.B.
Underwater Obstacles of
various Types are being
laid with great rapidity
and are likely to extend
along further stretches
of the coast.

MIKE

GREEN

HOUSES BRICKED UP AND LOOPHOLED

Les Roches de Ver

MINEFIELD
& ROWS DEEP

SEA WALL

la Rivière

SATURATED GROUND
LIABLE TO FLOODING

Mont Fleury

le Buisson

Vaux

Ver-sur-mer

Natural Scale
The Grid shown is the Lambert Zone 1.

Main Roads		Bridge	
Secondary Roads		Church	
Railway		Calvary	
Tramway		Light	
Tramway on roadside		Wooded Areas	
Spot Heights, Contours (in Metres)			

Scale of Yards
Scale of Metres

86 Gold Beach, Normandy, Admiralty Hydrographic Department, England, 1944

This map of Gold Beach, one of the five beaches selected for the D-Day landings in Normandy in June 1944, was produced by Royal Navy hydrographers working under the direction of Sir John Edghill. The beach was more than 5 miles (8 km) wide, with the small port of Arromanches—where it was planned to set up a prefabricated Mulberry harbor—at the western end. Preparing the map involved plotting the angle of slope of the beach plus as much as could be ascertained about the nature of the German defenses. On paper, these looked formidable, but in the event, following a devastating naval bombardment, German infantry resistance was less determined than had been expected, while the German armor at nearby Bayeux failed to reach the Allied beachhead. The landings started at 7.25 a.m. on June 6. By the evening, the British had landed 25,000 men, penetrated 6 miles (10 km) inland, and linked up with the Canadians advancing from Juno Beach to the left. Mapping such as this was vital to the success of the operation.

military undertaking in history was by no means an exaggeration. Maps played a crucial part in the operation's planning process. In fact, the Normandy landings might well have been impossible to accomplish had it not been for the covert hydrographic surveying that preceded the assault.

Accurate, up-to-date intelligence was naturally vital to the success of the attack. This meant that the information shown on the specially compiled maps of the landing beaches had to be gathered in the utmost secrecy without the Germans becoming aware of anything unusual occurring in the areas that were being surveyed. Photographic reconnaissance flights were of some help to the mapmakers, but the bulk of the task fell to parties of hydrographic surveyors sent out by the British Admiralty to make rapid reconnaissance surveys of the Normandy coast under cover of darkness. The beaches they had to survey were codenamed Gold, Juno, Sword, Omaha, and Utah.

The planners' intentions were for British and Canadian troops to land on the first three beaches, while Omaha and Utah were to be the preserve of the Americans. US Navy lieutenant William A. Bostick (b. 1913) was put in charge of mapping the latter. He and his team, many of whom had served with him when he had prepared his landing maps for the invasion of Sicily the previous year, used a variety of sources to prepare their D-Day maps. These included depth soundings and sketches taken and made by US Navy frogmen, panoramic aerial photographs of the shore taken by low-flying reconnaissance aircraft and, to strike a historical note, a nautical chart that, according to Bostick's later reminiscences, "dated from the time of Napoleon."

Operation Neptune

The task of mapping the British and Canadian beaches, a vital part of the aptly codenamed Operation Neptune, got under way in August 1943, when Royal Navy hydrographic surveyors based at Cowes on the Isle of Wight were issued with two shallow-draft landing craft from which to chart the Normandy coast east of Cherbourg. The low profiles of the landing craft would make it hard for German radar to spot them, while canoes would be used when landings on the actual beaches were required. The surveying could take place only on moonless nights when there was a high tide. On suitable nights, gunboats towed the landing craft halfway across the English Channel before the latter were cast off to proceed under their own power to the beach areas. The surveyors, who were able to start their work just before midnight, were under strict orders to quit the French coast by 4 a.m. to rendezvous with the gunboats and be towed safely home again.

The information the surveyors had so painstakingly gathered was added to the detailed charts and maps of the approaches to the invasion areas and the areas themselves that were being compiled by the Royal Navy's hydrographic department, based in Bath. Among the factors the draftsmen had to establish were the angle of slope of each beach and whether there were any sudden variations in it; their consistency, or rockiness; and the location and nature of the German off-shore and on-shore defenses. As well as existing maps and charts and the results of the covert surveys, the draftsmen had another tool at their disposal. Also in the summer of 1943, the BBC broadcast an appeal for peacetime holiday postcards and photographs of the French coast to be sent to the Admiralty. The response was amazing. More than 7 million photographs and postcards were received and the relevant ones were then passed on to the D-Day planners and cartographers.

Escape maps

One of the innovations that came into its own with the outbreak of World War II was the notion of printing maps on fabric to use as escape maps. From early on in the war, for instance, British bomber crews were issued with survival kits to help them evade capture or escape imprisonment should they be shot down over hostile territory. Typically, these kits included a tiny saw blade, a

RAF escape maps for sale, London, December 1945
Customers examine RAF escape maps for sale in a London department store as alternatives to traditional scarves. By the end of the war, some 1.3 million escape maps had been manufactured.

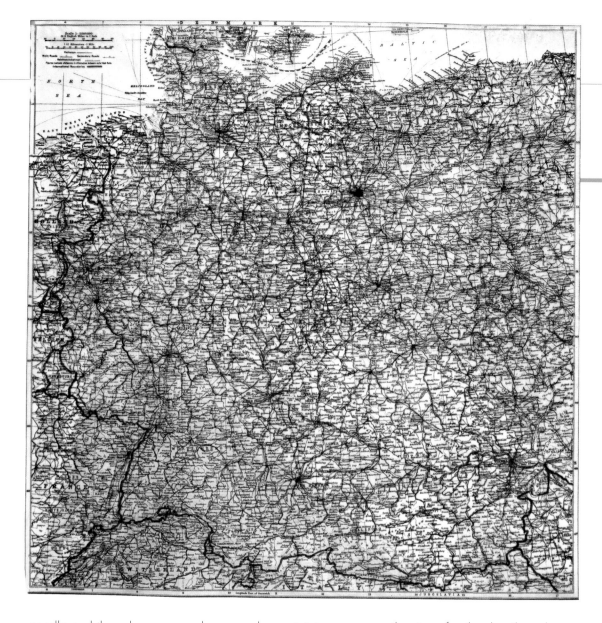

87 Silk Escape Map of
Germany, MI9,
England, 1939
Produced by British
military intelligence working with the
map publisher John Bartholomew and
the games maker Waddington's, silk
escape maps were first issued to RAF
aircrew in October 1939. The maps,
which were mainly small scale and
covered large areas, were usually
copies of existing Bartholomew
maps—the company generously
waived its royalties in the interests of
the war effort. Later on in the war, as
more and more Allied aircrew were
shot down and captured, POWs started
to make their own maps in their
prison camps. In at least one POW
camp near Brunswick, they even
managed to set up and run their own
printing press right under the noses of
their German guards.

needle and thread, currency, phrase cards, a miniature compass the size of a thumbnail, and
most crucially, a silk escape map.

There were several reasons for the choice of silk as the mapping medium. Fabrics such as silk
are extremely durable materials, which will not disintegrate in water and cannot be damaged by
repeated folding and unfolding. Most importantly in this particular context, maps on silk were easy
to conceal. Sewn into the lining of a jacket, or hidden in a cigarette tin or the hollowed-out heel
of a boot, such a map was unlikely to be found, at least in a cursory initial search. In Britain, the
task of producing these maps was the responsibility of MI9, the escape and evasion branch of the
intelligence service. The cartography came from Bartholomew's, the great atlas makers, while
Waddington's, best known in peacetime for producing such games as Monopoly, carried out the
manufacturing. The maps themselves were often double-sided. They were extremely detailed,
showing features such as cities, towns, villages, lakes, rivers, roads, railroads, and mountain
passes. Sometimes they included coverage of ocean currents and navigational star charts.

In the USA, MIS-X, the American counterpart to MI9, was quick to adopt the idea. From 1942
onward, escape maps became standard issue for all US flying personnel in all theaters of war.
Soon, they were being issued to paratroopers as well as to aircrew, while many were being
smuggled into POW (prisoner of war) camps concealed in Red Cross aid parcels.

Some 1.3 million escape maps had been produced by the end of World War II. It has been
estimated that, out of the 35,000 Allied troops who managed to escape from behind enemy
lines, more than half used a clandestine map to help them on their way.

MAPS AS PROPAGANDA

German propagandists during and after World War I took full advantage of the potential of cartography when it came to devising a suitably biased representation of reality and, from 1933 onward, the Nazis put this power to greatest use. It was also the Germans who coined the two infamous terms *Lebensraum* (living space) and *Geopolitik* (geopolitics).

As early as 1871, the term *Lebensraum* was a popular political slogan during the establishment of a united Germany. At this time, *Lebensraum* usually meant finding additional "living space" by adding colonies, following the examples of the British and French empires. Hitler changed the concept of *Lebensraum*. Rather than adding colonies to make Germany larger, Hitler wanted to enlarge Germany within Europe. The term *Geopolitik* was originated by Rudolf Kjéllen, a Swedish political scientist, in 1905. As a sub-branch of political geography, geopolitics focused on the spatial development and needs of the state. In the 1920s, German geographer Karl Haushofer used *Geopolitik* to support German expansion. Haushofer believed that densely populated countries such as Germany should be entitled to expand and acquire the territory of less populated countries.

Cartography in Germany was increasingly controlled to serve the purposes of the Nazi government. Between July 1934 and June 1944, some 60 map-related regulations were issued by the German government dictating what could and could not be shown on maps and how best to show it. Cartographic control tightened still further in 1937, when a law gave the Ministry of the Interior authority over all private mapmakers, who now had to conform to official policy on geographical names and the colors used on maps.

From that point on, all German maps sang the same tune. For areas lost to Germany in 1919 by the terms of the Treaty of Versailles, such as the twin provinces of Alsace and Lorraine, which the treaty returned to France, it was decreed that "the German names are to be printed in a bold type, the other [local names] in smaller type and in parenthesis." The *Saar-Atlas* of 1934 supported the German case for the reintegration of the Saar in the fatherland. The linguistic map on its opening pages showed both the Saar and Alsace as being German. On world political maps, it was now Greater Germany—not the British Empire—that was to be shown in red.

Nazi triumphs, such as the *Anschluss* (union) with Austria in 1938, were also celebrated in map form. One such map, issued shortly after the event, showed the two countries combined into a single landmass with the head of the Führer stamped firmly in the center. The notion behind the map was to stress the idea of a single, united Germanic people finally coming together to fulfill its manifest historical destiny in a 1000-year Reich that would dominate Western civilization. More mundanely, it was also intended to influence voting in Austria in favor of the *Anschluss* in the plebiscite held immediately after the Nazi takeover of power.

88 Austrian *Anschluss*, Germany, 1938

Printed on card for easy mass circulation, this map commemorates the *Anschluss* (union) of March 1938, as a result of which Austria became part of the Third Reich. Hitler, as architect of the annexation, dominates the map, whose message is blunt: *Ein Volk, Ein Reich, Ein Führer* ("One people, one fatherland, one leader"). It is obvious from the map what the next Nazi targets might be: the part of Czechoslovakia that jutted into the Reich and the so-called Polish Corridor that separated East Prussia from Germany proper. The latter had been created by the Treaty of Versailles to give the newly independent Poland access to the Baltic and was always a potential bone of contention.

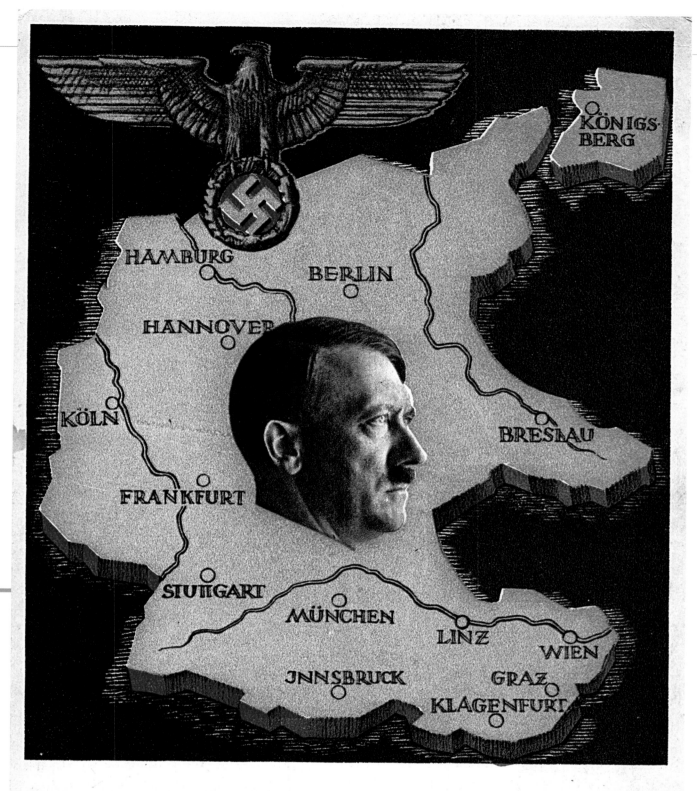

یہ سپاہی ہندوستان کی حفاظت کر رہا ہے ۔ وہ اپنے گھر اور گھر والوں کی حفاظت کر رہا ہے ۔
اپنے گھر والوں کی مدد کرنے کا سب سے اچھا طریقہ یہ ہے کہ فوج میں بھرتی ہو جاؤ ۔

84

THE TIMES PRESS BOMBAY

89 Indian Army Recruiting Poster, India, 1914

Rifle at the ready and bayonet fixed, an Indian soldier stands guard protecting the motherland against external threats at the start of World War I. As in the case of other recruiting posters published in India at the time, the idea was to urge recruits to come forward to defend their homeland, rather than to fight for Great Britain itself. The campaign was not universally successful. Although the Indian Army increased its numbers from a pre-war strength of 155,000 to 573,000 by the time of the armistice in 1918, the officer corps remained overwhelmingly British and the rank-and-file Indian. While Indians were fighting in Mesopotamia and on the Western Front during World War I, the first demands for Home Rule were being raised by the founders of the Indian National Congress.

When it came to demographic maps, the Nazis twisted statistics in their quest to legitimize their claims to vast tracts of land in Eastern Europe, notably Czechoslovakia, Poland, and the Ukraine. The maps they produced were intended to demonstrate the "Germanic nature" of these regions. Ultimately, the information was used to identify and locate those who were deemed "unfit" to live in the Reich and so contributed to the efficiency of Nazi genocide.

Distorting cartography

The Germans were not alone in appreciating the power of modern cartography and the possibilities of its exploitation and manipulation. In 1919, the Czechs published a linguistic map of Central Europe emphasizing where Czech speakers were located in bright scarlet. Hungary produced a similar map in 1923 in support of its claim for the return of the territories it had lost as a result of the Treaty of Trianon three years earlier, while in 1937 Ukrainian nationalists living in exile in Poland produced an atlas of their homeland and the territories surrounding it to bolster Ukrainian national identity in the face of oppressive Soviet rule.

Map colors and projections could be subtly manipulated to change the physical dimensions of a nation, or to make a country appear to be vulnerable or secure, dangerous or feeble. During the so-called period of "phoney war" in the fall and winter of 1939, for instance, the French government sanctioned the publication of a map that deliberately distorted the size of the British and French Empires as opposed to the territories controlled by the Third Reich, the map being accompanied by the slogan "We shall win because we are the stronger." Despite the best intentions of the map's creators, the result did little to boost morale in France at the time and was swiftly forgotten after Hitler's armies struck in the west in May 1940.

The propaganda map that had been produced to rally India behind the British Raj during World War I was equally ineffective. The idea was that, by showing a stalwart Indian Army infantryman standing in defense of his own country, latent feelings of loyalty toward Britain would be encouraged. However, the map could equally well be seen as endorsing the demands of the emerging Indian nationalists, who, with the launch of their Home Rule for India campaign, were deeply opposed to the continuance of British rule in its traditional form.

Redrawing the map

Sometimes entire countries have disappeared from the map, the classic example being the way in which Poland disappeared after its third and final partition between Russia, Prussia, and Austria at the end of the 18th century. It was not to regain its place on the map of Europe until the Treaty of Versailles recreated it as an independent state after World War I.

Of all the countries in the world, however, no nation's map has been redrawn as often and as radically as that of Israel. The roots of the process that eventually culminated in the establishment of the modern Israeli state can be traced back to 1917, when the British government issued the Balfour Declaration, in which it declared its support for "a national home for the Jewish people" in Palestine.

As early as 1935, the Peel Commission, appointed by the British government to devise a policy for Palestine—since the end of World War I it had been a British protectorate—had recommended the partition of the country into a small Jewish state around Tel Aviv and a much larger Arab one. It was the Arabs and not the Jews who rejected the scheme. Then, in 1947, after guerrilla war had broken out between the British garrison and Jewish underground resistance organizations, the problem was handed over to the United Nations. The result was another partition plan. Palestine was to be divided into two independent states—one for the Palestinian Arabs and another Jewish—with Jerusalem internationalized. In the event, the settlement failed to win acceptance from either side. Immediately following the British withdrawal, as Israel proclaimed its independence, its Arab neighbors attacked, determined to drive the Jews into the sea. Paradoxically, as a result of the war, the Israelis ended up with much of the territory the UN had assigned to the Arabs. More than 5 million Palestinians were displaced.

Cartographers, of course, have been deeply involved in mapping this continuing problem and the various plans suggested for its solution. Starting with the 1949 territorial settlement—the map used as the basis for this was the 1944 British Survey of Palestine—which marked the conclusion of an armistice between Israel and its neighbors, the country has gone through immense territorial changes as a result of the continuing conflicts in which it has been embroiled.

Following instructions

Maps vary in what they show, depending on the viewpoint and purpose of their creators—or, rather, those of the political masters whose instructions the cartographers are following. Well into the 20th century, for instance, Africa was perceived as a continent practically without history, and what little it had was held to be the direct result of European colonization rather than any indigenous presence. All of the colonial powers used mapmakers as propagandists to exaggerate the extent of their territorial claims on the continent.

There are many similar examples that could be cited, from throughout the history of cartography. Today, Indian maps, for instance, show the disputed territory of Kashmir as lying within the boundaries of India, while Pakistani mapmakers show it as part of Pakistan. Similarly, in the period prior to the outbreak of the Falklands War in 1982, Argentine maps showed the disputed islands—Las Malvinas, as they were named—as an integral part of the mother country, even though they were internationally recognized as a British colony.

90 **Palestine, British Survey of Palestine, 1949**
This map was originally prepared in 1944 by the British Survey of Palestine and was then used by negotiators as they attempted to settle the Israeli borders. The map is detailed enough to show clearly the names of many towns and villages that figured in the Israeli battle against the Arabs to preserve their independence in 1948, as well as many Palestinian settlements that either disappeared or became Jewish ones as a result of the 1949 armistice. By its terms, the west bank of the River Jordan and the Gaza strip became distinct geographical units, a situation that lasted until the 1967 Six Days War brought them both under Israeli control.

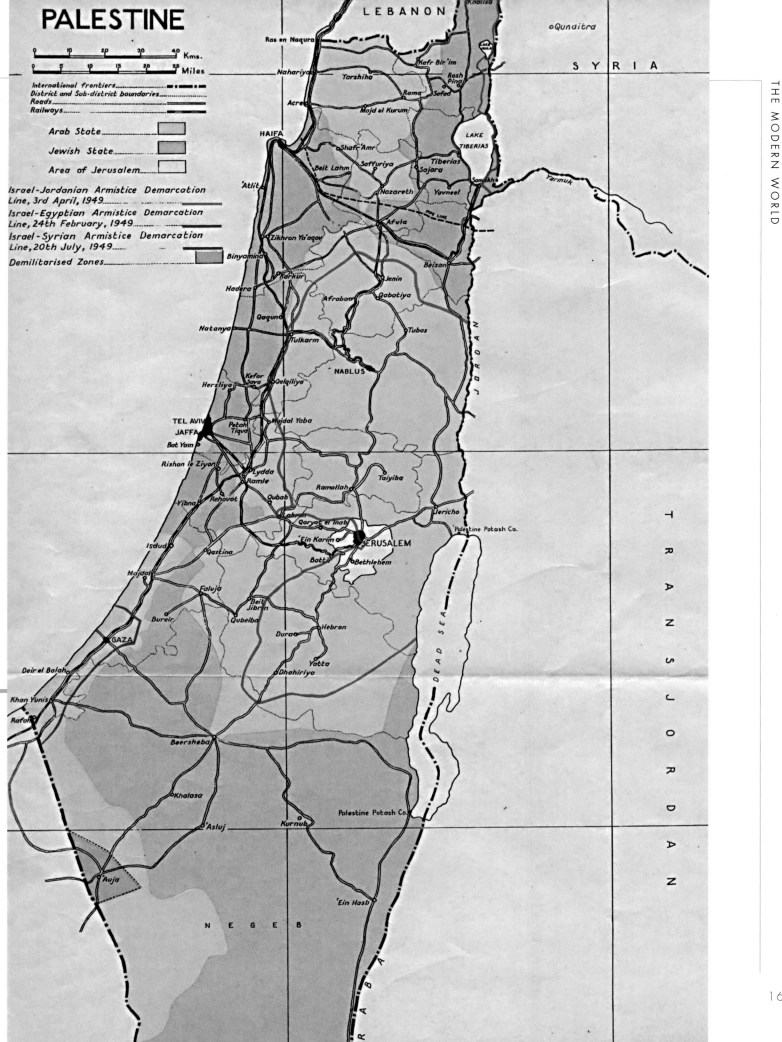

PALESTINE

Kms.
Miles

International frontiers ....
District and Sub-district boundaries ....
Roads ....
Railways ....

Arab State ....

Jewish State ....

Area of Jerusalem ....

Israel-Jordanian Armistice Demarcation
Line, 3rd April, 1949 ....

Israel-Egyptian Armistice Demarcation
Line, 24th February, 1949 ....

Israel-Syrian Armistice Demarcation
Line, 20th July, 1949 ....

Demilitarised Zones ....

LEBANON
oQunaitra
SYRIA

Ras en Naqura
Khalisa
Nahariya
Tarshiha
Kafr Bir'im
Rosh Pina
Rama
Safed
Acre
Majd el Kurum
LAKE
HULA

HAIFA
Shafr 'Amr
Beit Lahm
Saffuriya
Tiberias
LAKE
TIBERIAS
'Atlit
Nazareth
Sajara
Samakh
Yarmuk
Afula
Yovneel
Pipe Line

Zikhron Ya'aqov
Beisan
Binyamina
Karkur
Jenin
Hadera
Afrobao
Qabatiya
Qaquna
Natanya
Tubas
Tulkarm
Kefar Sava
Herzliya
Qalqiliya
NABLUS
TEL AVIV
JAFFA
Petah Tiqva
Majdal Yaba
Bat Yam
Rishon le Ziyon
Lydda
Ramle
Taiyiba
Yibna
Rehovot
Qubab
Ramallah
Isdud
Qastina
Lahrun
Qaryat el Inab
Jericho
Palestine Potash Co.
Majdal
'Ein Karim
JERUSALEM
Faluja
Batti
Bethlehem
Beit Jibrin
Bureir
Qubeiba
Dura
Hebron
GAZA
Yatta
Deir el Balah
Dhahiriya
Khan Yunis
Rafah
DEAD SEA
Beersheba
Khalasa
TRANSJORDAN
'Asluj
Kurnub
Palestine Potash Co.
'Auja
'Ein Hasb
N E G E B
A R A B A

JORDAN

165

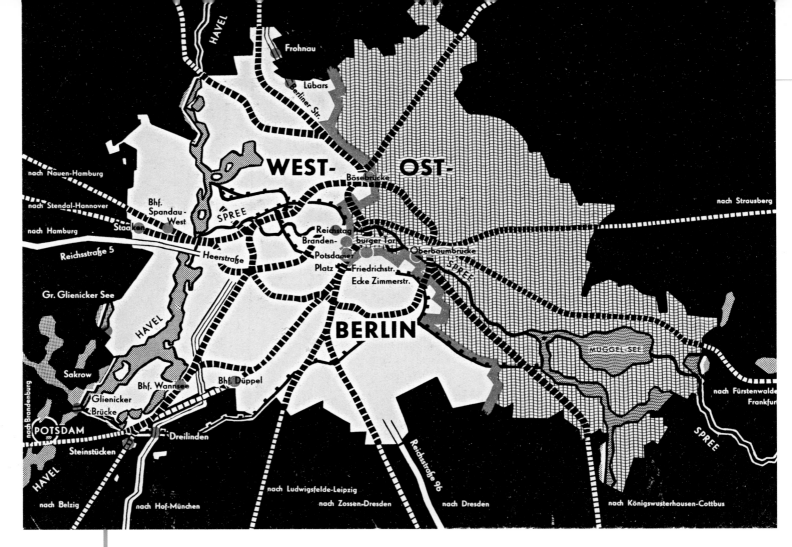

91

East and West Berlin, West Germany, 1959
A divided city since 1945, the former capital of the German Reich was physically split in two in 1961, when, fearful of the increasing numbers of its citizens who were fleeing to the west, the East German Communist regime closed the border between the two halves of the city by erecting the Berlin Wall. Drawn in 1959 for a West German government pamphlet, this map was emotively entitled "Free City between Barbed Wire." Such mapping concreted the divide in the public imagination, even before the building of the wall. For almost 30 years until its demolition in 1989, the wall was a symbol of the Cold War. This ended only with Communism's collapse, the fall of the wall, and the reunification of Germany a year later.

THE COLD WAR

After the founding of the USSR, maps were soon recognized as a powerful means of presenting both the triumphs of Communism and promoting the future of the Soviet cause. In 1928, for instance, the Division of Military Literature of the State Publishing House of the Red Proletariat produced a fascinating series of ten maps to illustrate the sequence of events from the Bolshevik Revolution of October 1917, the intervention by the Allies on behalf of the White Russians and the subsequent civil war to the eventual destruction of the counter-revolutionary forces by the triumphant Red Army. With their bold use of colors and striking employment of symbols, such maps brought the struggle between Communism and capitalism to life. At that time, illiteracy was common in the Soviet Union, so such maps were an excellent means of getting the Soviet propaganda message across to the widest possible audience.

After World War II, the Soviet dictator Josef Stalin (1879–1953) instituted a major national mapping program, which involved the detailed resurveying of the entire country with its newly expanded boundaries. After the project was completed in 1954, the year after Stalin's death, Soviet mapmakers turned their attention to many countries that were seen as potential adversaries in the Cold War, including the USA. Meanwhile, Soviet maps of the USSR and its satellites were secret unless they were of very small scale or were highly simplified and devoid of sensitive details. For its part, the USA was equally aware of the potential of maps as vehicles for propaganda. Indeed, the US government favored the employment of map projections that made the USSR look larger than it actually was, and so more of a threat to the "free world." On the other hand, Soviet mapmakers frequently depicted the USSR as being encircled by aggressive Western militarism.

At the height of the Cold War, what might best be termed journalistic cartography played a potent role in influencing the American view of what was then popularly termed the "Red Menace." The great magazines of the period, such as *Time, Life*, and the *Saturday Evening Post*, routinely used maps to illustrate the major stories they featured dealing with Cold War issues. These maps set out to be visually eye-catching. Equally, they were almost always biased.

As early as April 1946, just a month after Winston Churchill had warned of the descent of "an iron curtain" across Europe, *Time* magazine featured a map titled *Communist Contagion*. On it, a vividly red-colored USSR spreads the contagion of Communism. The USSR's neighbors, and other states, are categorized according to how much the mapmakers considered them to be at risk: they were exposed, quarantined, or infected. Another map, entitled Europe from Moscow, which appeared in *Time* in 1952, similarly depicted a blood-red flood tide of Communism seeping out from its Soviet homeland and threatening to engulf the whole of the European continent. Advertisements, such as one for the Grumman Aircraft Company in the January 5 1953 issue of *Life* magazine, often included maps as a visual way of highlighting the threat of Communist expansion.

Such images were powerful enough to linger in the collective American folk memory long after they were conceived, even in the changed political climate of the last decades of the 20th century. As late as 1986, the American artist Andy Warhol (1928–87) created a map of purported Soviet missile bases. This turned out to be one of the last paintings he was able to execute before his death. It is generally agreed that, in a striking throwback to some of Warhol's earliest work, the black-and-white image, which was hand-drawn and then silk-screened, was a direct copy of an original newspaper map, but the question of why Warhol chose this particular subject in the first place is one that remains unanswered.

It may be that the map was intended as an act of subversion—with its sketchy, inaccurate data, it certainly served no utilitarian purpose—but it seems more likely that Warhol was highlighting what he believed to be the ephemeral nature of information and the banality of contemporary consumer culture. His target, therefore, might well have been the instant graphics of the type favored by the newspapers and news magazines of the time and the false air of authority they exuded. An analysis of news maps in German newspapers at the time of the 1979 Soviet invasion of Afghanistan, for instance, showed that 60 percent of them came without a scale, while only 22 percent gave sources for the data they were portraying.

CITY LIFE

In the 20th century, map use grew and grew as the public became accustomed to consulting maps and plans on an everyday basis, both for work and for leisure purposes. Many new variants of map, such as the one draftsman Harry Beck (1903–74) created of the London Underground system between 1931 and 1933, were the result (see p.170). Their prime aim was to satisfy a growing popular need.

The reason Beck was asked to produce his new map was simple. By the 1930s, London's subway network

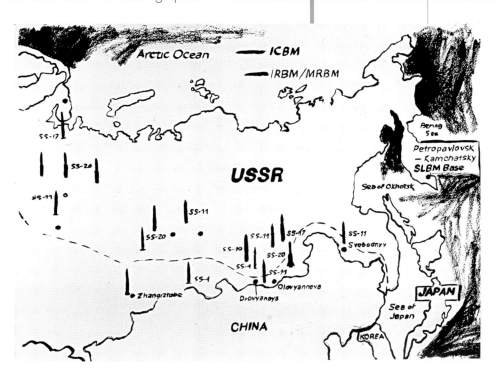

had expanded to the point where it had become difficult to visualize all the new stations and lines within a conventional geographical format. Passengers complained that the existing maps, which tried to ease travel on the underground network as well as to present the accurate geographical location of the stations above ground, were crowded, confusing, and generally hard to read.

Beck's answer, inspired by his knowledge of electrical circuitry, was to abandon any notion of scale and to depict the various lines as being more or less straight verticals, horizontals, and diagonals. The central area was enlarged for greater legibility, while the outlying suburban areas, with fewer stations, were compressed. This also had the benefit, in a period of dramatic growth and new building for London, of making it appear as though the suburbs were closer to the heart of the city than they were in reality. Interchanges between the various lines and overground rail services were clearly marked with diamond symbols.

The map Beck devised—he continued to develop and refine it until his retirement in 1959, though the basic design remained the same—is more of a diagrammatic plan than a conventional cartographic rendering. Yet it was a utilitarian response to a decidedly 20th-century problem. The map is now universally considered to be an icon of 20th-century graphic design. Since its creation, it has influenced subsequent maps of subway systems all over the world in places as far removed as New York, St. Petersburg, and Sydney.

However, Beck's approach did not meet with universal approval at the time. His conservative chiefs at London Transport had to be persuaded to accept his bold ideas, since their immediate reaction was to dismiss his map because it was not like the purely geographical maps that had been used to show the network before. Later in the 1930s, he was asked to devise a new map of the Paris Métro, but although he continued to work on drafts of it up until 1951, the French transport authorities rejected the result, considering the treatment inappropriate for their network.

93 Westminster, *A-Z Street Atlas*, Phyllis Pearsall, England, 1963

One of the most influential city street directories of the 20th century came into existence almost by accident. Phyllis Pearsall, its originator, was not a professional mapmaker. Rather, she was inspired to create her revolutionary street atlas by her failure to locate a street she was looking for in Belgravia on the Ordnance Survey map of the area. It was this that persuaded her to devise a more efficient way of helping people to navigate the labyrinthine city streets. Methodically, she divided the city into sections, personally walking the streets of each in turn. These were linked to a comprehensive index, one of the keys to the guide's success when it was first published in 1936. This map is from the 1963 edition.

London from A to Z

While Beck's map was making it easier for Londoners and visitors to the city to find their way around underground, Phyllis Pearsall (1906–96) was performing a similar service for travelers on the surface. The A-Z street atlas of the city she conceived, compiled, and designed remains one of the most ingenious and best-known pieces of information design produced in the first half of the 20th century. Pearsall was not a trained cartographer: she was a writer and a painter. She was inspired to start her mammoth task in 1935 when, planning to attend a party and armed with the most up-to-date street map of London, she failed to find the street of the house where it was being held. Working up to 18 hours a day, she walked 3,000 miles (4,800 km) in order to map all of

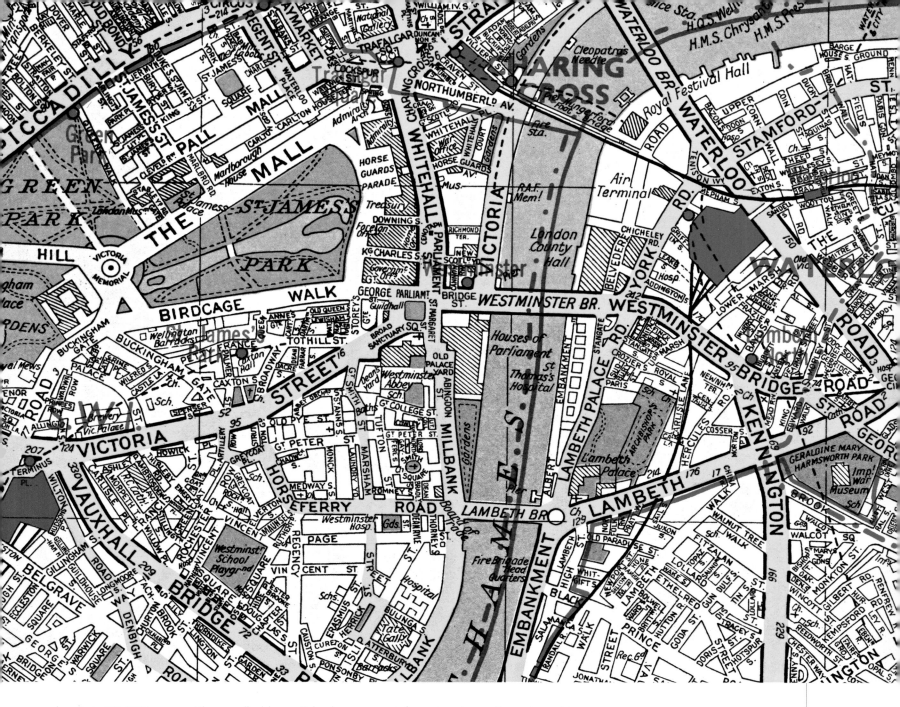

the city's 23,000 streets. She recalled later: "I had to get my information by walking. I would go down one street, find three more and have no idea where I was." It was perhaps not surprising that errors crept in during the compilation process. It was not until the final checks were being made as the atlas was being readied for printing, for instance, that Pearsall realized she had omitted to include Trafalgar Square.

Because of the sheer size of the city, Pearsall decided against trying to produce a sheet map of it. By its very nature, this would have been extremely cumbersome to use. It would also have been very difficult to read, since she would have been forced to map at an extremely small scale. Instead, she decided to divide her maps into sections, each of which would be keyed to a comprehensive street index.

It was this all-important index that gave Pearsall the idea for the atlas's name. The atlas appeared in 1936, published by a company that Pearsall had set up herself in the face of the atlas's rejection by the established cartographic publishers to which it had been submitted. The publication was a truly Herculean effort, made all the more so since the company she founded initially had a staff of two: Pearsall herself and a single draftsman.

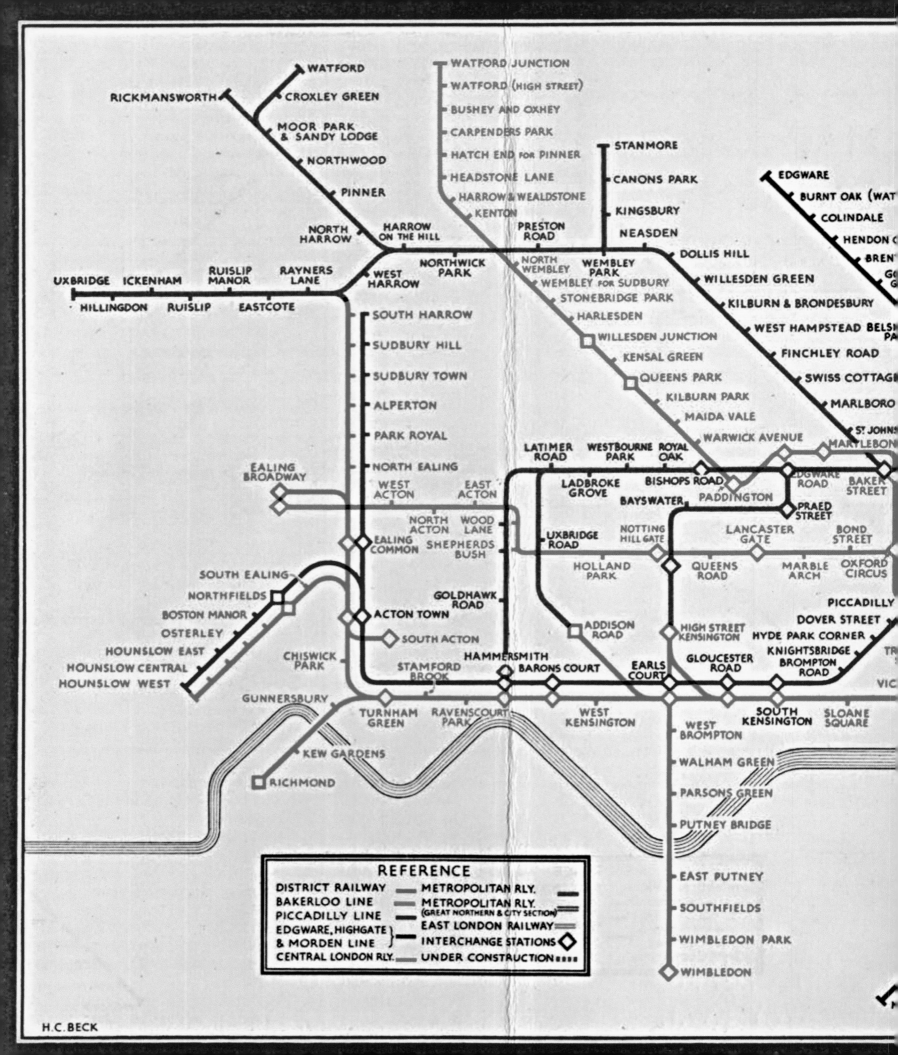

WATFORD
RICKMANSWORTH
CROXLEY GREEN
MOOR PARK & SANDY LODGE
NORTHWOOD
PINNER

WATFORD JUNCTION
WATFORD (HIGH STREET)
BUSHEY AND OXHEY
CARPENDERS PARK
HATCH END for PINNER
HEADSTONE LANE
HARROW & WEALDSTONE
KENTON

STANMORE
CANONS PARK
KINGSBURY
NEASDEN

EDGWARE
BURNT OAK (WAT
COLINDALE
HENDON
BREN

NORTH HARROW
HARROW ON THE HILL
WEST HARROW
NORTHWICK PARK
PRESTON ROAD
NORTH WEMBLEY
WEMBLEY PARK
DOLLIS HILL
WEMBLEY for SUDBURY
STONEBRIDGE PARK
WILLESDEN GREEN
KILBURN & BRONDESBURY

UXBRIDGE   ICKENHAM   RUISLIP MANOR   RAYNERS LANE
HILLINGDON   RUISLIP   EASTCOTE

SOUTH HARROW
SUDBURY HILL
SUDBURY TOWN
ALPERTON
PARK ROYAL
NORTH EALING

HARLESDEN
WILLESDEN JUNCTION
KENSAL GREEN
QUEENS PARK
KILBURN PARK
MAIDA VALE
WARWICK AVENUE

WEST HAMPSTEAD   BELSH PA
FINCHLEY ROAD
SWISS COTTAGE
MARLBORO
S? JOHNS
MARYLEBON

EALING BROADWAY
WEST ACTON   EAST ACTON
LATIMER ROAD   WESTBOURNE PARK   ROYAL OAK
LADBROKE GROVE   BISHOPS ROAD
BAYSWATER   PADDINGTON
EDGWARE ROAD   BAKER STREET
PRAED STREET

NORTH ACTON   WOOD LANE
EALING COMMON   SHEPHERDS BUSH
UXBRIDGE ROAD   NOTTING HILL GATE
LANCASTER GATE   BOND STREET

SOUTH EALING
NORTHFIELDS
BOSTON MANOR
OSTERLEY
HOUNSLOW EAST
HOUNSLOW CENTRAL
HOUNSLOW WEST

HOLLAND PARK   QUEENS ROAD   MARBLE ARCH   OXFORD CIRCUS

GOLDHAWK ROAD
ACTON TOWN
SOUTH ACTON
CHISWICK PARK
STAMFORD BROOK
HAMMERSMITH   BARONS COURT
ADDISON ROAD
HIGH STREET KENSINGTON
EARLS COURT
GLOUCESTER ROAD

PICCADILLY
DOVER STREET
HYDE PARK CORNER
KNIGHTSBRIDGE BROMPTON ROAD
TR

GUNNERSBURY
TURNHAM GREEN
RAVENSCOURT PARK
WEST KENSINGTON
WEST BROMPTON
SOUTH KENSINGTON
SLOANE SQUARE
VIC

KEW GARDENS
RICHMOND
WALHAM GREEN
PARSONS GREEN
PUTNEY BRIDGE

EAST PUTNEY
SOUTHFIELDS
WIMBLEDON PARK
WIMBLEDON

REFERENCE

DISTRICT RAILWAY	METROPOLITAN RLY.
BAKERLOO LINE	METROPOLITAN RLY. (GREAT NORTHERN & CITY SECTION)
PICCADILLY LINE	EAST LONDON RAILWAY
EDGWARE, HIGHGATE & MORDEN LINE	INTERCHANGE STATIONS ◇
CENTRAL LONDON RLY.	UNDER CONSTRUCTION ▪▪▪▪

H.C. BECK

COCKFOSTERS

OPEN
MIDSUMMER
1933

ENFIELD WEST

SOUTHGATE

ARNOS GROVE

BOUNDS GREEN

WOOD GREEN

TURNPIKE LANE

MANOR HOUSE

HIGHGATE

TUFNELL
PARK

KENTISH
TOWN

FINSBURY PARK

ARSENAL
(HIGHBURY HILL)

DRAYTON PARK

CAMDEN
TOWN

HOLLOWAY
ROAD

HIGHBURY & ISLINGTON

MORNINGTON
CRESCENT

CALEDONIAN
ROAD

CANONBURY & ESSEX ROAD

KINGS CROSS
ST. PANCRAS

USTON

OLD STREET

EUSTON
SQUARE

ANGEL

WARREN
STREET

FARRINGDON ALDERSGATE

MOORGATE

GOODGE
STREET

RUSSELL
SQUARE

CHANCERY POST
LANE OFFICE

LIVERPOOL STREET

HOLBORN

BANK

BRITISH
MUSEUM

ALDWYCH

SHOREDITCH

COVENT
GARDEN

MANSION
HOUSE

ALDGATE

TO
BOW ROAD
BROMLEY
WEST HAM
PLAISTOW
UPTON PARK
EAST HAM
BARKING
UPNEY
BECONTREE
HEATHWAY
DAGENHAM
HORNCHURCH
UPMINSTER
& SOUTHEND

MONUMENT

STEPNEY
GREEN

RE

STRAND

BLACKFRIARS

CANNON
STREET

MARK
LANE

ST. MARYS

WHITECHAPEL

MILE
END

TEMPLE

ALDGATE
EAST

CHARING
CROSS

SHADWELL

STER

LONDON BRIDGE

WAPPING

WATERLOO

LAMBETH
NORTH

BOROUGH

ES

ELEPHANT
& CASTLE

ROTHERHITHE

SURREY DOCKS

KENNINGTON

OVAL

STOCKWELL

CLAPHAM NORTH

NEW CROSS
GATE

NEW
CROSS

CLAPHAM COMMON

CLAPHAM SOUTH

BALHAM

TRINITY ROAD (TOOTING BEC)

TING BROADWAY

UNDERGROUND

S WOOD

BLEDON (MERTON)

94 London
Underground,
Harry Beck,
England, 1935
Electrical engineer turned
draftsman Harry Beck
abandoned geographically
accurate cartography when he
produced his striking plan of
London's burgeoning subway
system in 1933, the year in
which London Transport was
founded. Beck humorously
stated that he had been
inspired by electrical circuitry
in formalizing his plan, and
the result was a striking
schematic. The use of colors
for distinguishing different
lines, and the inclusion of just
one aboveground feature—the
River Thames—made the map
instantly understandable.
The simplicity and clarity
of the map's approach set
the standard for all
subsequent subway maps
around the world. This is the
1935 card folder version
issued to the public.

Another dimension

Traditionally, 20th-century city maps were confined to providing a picture of life at ground level. They were organized street by street with blocks for buildings—and the occasional green space—filling up the gaps between streets. Then, in the late 20th century, computer technology made it possible to produce three-dimensional axonometric maps that transformed the way in which we view cities and the cartographic pictures we have of them.

A wide range of data is needed in order to create such maps. These data include architectural plans of the buildings that are to feature on the map in question, aerial and ground-level photographs, satellite and stereographic images, topographic maps, laser scans, and what are termed Digital Elevation Models (DEWS). The data are digitized, combined, and electronically manipulated to create isometric views of the cities or areas of the cities it represents. Cartographers can manipulate the data to generate maps that show different combinations of information, from buildings and power lines down to individual trees and automobiles on the street.

By eliminating perspective, all the buildings represented remain in proportion to one another, while the true-to-life detail such maps capture is visually remarkable. The results serve a dual purpose. As well as being informative—many are produced for commercial and industrial purposes—they are more often than not striking works of art in their own right. Historically, they can trace their origins back as far as the bird's-eye-view representations of major European cities that started to appear from the 1500s onward, and the panoramic maps that became popular in the USA in the 19th century.

NEW PROJECTIONS

The changing political and social climate of the second half of the 20th century had a major impact on the course of cartography. The last few decades of the century saw determined efforts being made to replace the standard Mercator projection and other traditional projections, such as the Van der Grinten projection—invented in 1898 and used by the US National Geographic Society from 1922 to 1988—with new ones that were intended to be more representational. One of the most controversial of these was the Peters projection, named after Dr. Arno Peters (1916–2002), a German historian and cartographer, who devised it in the early 1970s (see p.174–5).

The Peters projection

Peters was a Marxist. His argument, which he first put forward in 1973, was that the projection he claimed to have devised—critics claimed that Peters, in fact, had based his work on a projection created by the Scotsman James Gall (1808–95) more than a century before—was far more realistic than Mercator's Eurocentric view of the world. Peters believed that the end of European colonialism and the coincidental rise of modern advanced technology meant that cartography had to move with the times. Mapmaking needed to be clear, readily understandable, and, above all, to free itself from traditional Western constraints and preconceptions.

Accordingly, Peters created what mapmakers term an "area accurate" view. Unlike maps employing the Mercator projection, such maps do not inflate the size of regions the further they are from the equator. Maps using the Peters projection, it was claimed, represented the sizes of all countries correctly, thereby doing justice to countries in the less developed regions of the world.

The increased size of nations in the developing world was one of the main reasons for the initial success of the Peters projection. At the time it was put forward, it was hailed by many as a

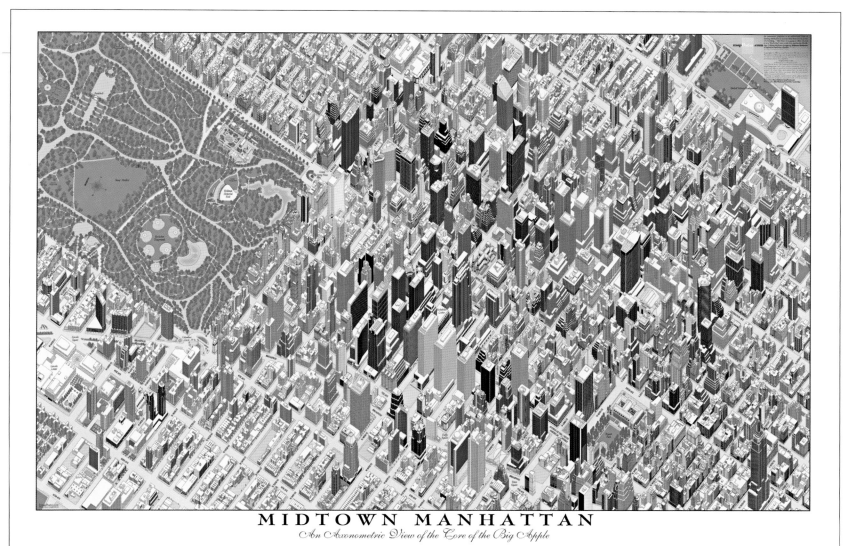

MIDTOWN MANHATTAN
An Axonometric View of the Core of the Big Apple

timely breakthrough that put an end to a traditional outdated way of representing the world that totally failed to reflect the changes that were taking place in global society. However, the Peters projection contains biases of its own. The lengths of South America and Africa are exaggerated, distorting the shape of the two continents, while the Arctic coasts of Canada and Russia are overlong.

World projections

In fact, there is only one "realistic" world projection: the orthographic projection, which shows a hemisphere just as you would see it from afar. This requires two circles in order to represent the whole Earth. All other projections have to sacrifice or distort some property, though cartographers aim to minimize all kinds of distortion. In 1923, J. Paul Goode (1862–1932) of the University of Chicago merged two existing projections—one covering the higher latitudes and the other the lower—to produce an equal-area projection that was far closer to global reality than the later Peters' attempt.

The Robinson projection, which the National Geographic Society selected to replace the Van der Gritten one in 1988, provides a far more accurate worldview that do Peters-based maps as well as being a vast improvement on its predecessor. Devised by Professor Arthur A. Robinson (1915–2004) of the University of Wisconsin in 1963, the projection was originally commissioned by the map and atlas publishers Rand McNally. Robinson called it the orthophonic ("right appearing") projection, but the name never caught on. It makes the world look flatter and squatter. While it does not eliminate geographic distortion entirely, it keeps it to a relatively low level over most of the world. As a result, maps using the Robinson projection are generally far more topographically accurate than their predecessors. In the first maps produced using it, the USSR, for

95
Axonometric Map of Midtown Manhattan, Ludington Ltd., USA, 2000

Using ground and aerial photography plus what are termed footprints of individual buildings as core reference, this three-dimensional map is one of a series of revolutionary views of the centers of major American cities. It is a stunning rendering of the heart of New York's business district, in which the Empire State Building, the Chrysler Building, the United Nations, Lincoln Center, and Central Park are all clearly distinguishable. The result is an informative map that at the same time is an outstanding work of art.

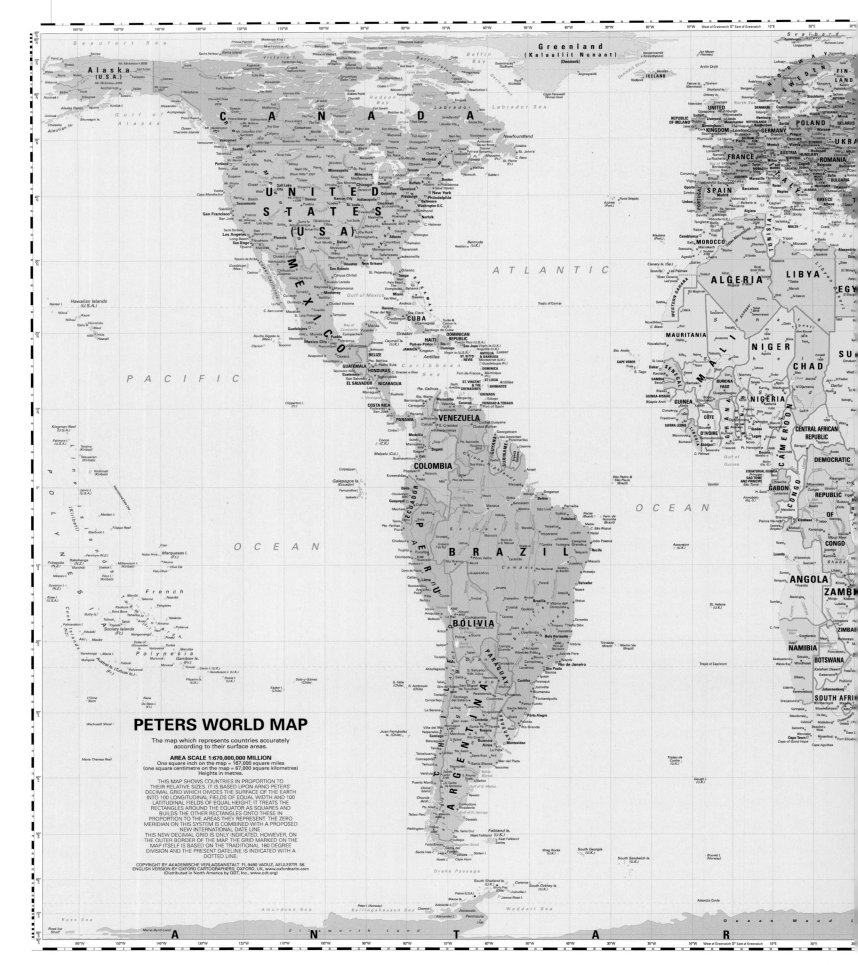

PETERS WORLD MAP

The map which represents countries accurately according to their surface areas.

AREA SCALE 1:670,000,000 MILLION
One square inch on the map = 167,000 square miles
(one square centimetre on the map = 67,000 square kilometres)
Heights in metres

THIS MAP SHOWS COUNTRIES IN PROPORTION TO
THEIR RELATIVE SIZES. IT IS BASED UPON ARNO PETERS'
DECIMAL GRID WHICH DIVIDES THE SURFACE OF THE EARTH
INTO 100 LONGITUDINAL FIELDS OF EQUAL WIDTH AND 100
LATITUDINAL FIELDS OF EQUAL HEIGHT; IT TREATS THE
RECTANGLES AROUND THE EQUATOR AS SQUARES AND
BUILDS THE OTHER RECTANGLES ONTO THESE IN
PROPORTION TO THE AREAS THEY REPRESENT. THE ZERO
MERIDIAN ON THIS SYSTEM IS COMBINED WITH A PROPOSED
NEW INTERNATIONAL DATE LINE.
THIS NEW DECIMAL GRID IS ONLY INDICATED, HOWEVER, ON
THE OUTER BORDER OF THE MAP, THE GRID MARKED ON THE
MAP ITSELF IS BASED ON THE TRADITIONAL 180 DEGREE
DIVISION AND THE PRESENT DATELINE IS INDICATED WITH A
DOTTED LINE.

COPYRIGHT BY AKADEMISCHE VERLAGSANSTALT FL-9490 VADUZ, AEULESTR. 56.
ENGLISH VERSION BY OXFORD CARTOGRAPHERS, OXFORD, UK, www.oxfordcarto.com
(Distributed in North America by ODT, Inc., www.odt.org)

96

Peters Projection World Map, Arno Peters, Germany, 1974

Devised by German historian and cartographer Dr. Arno Peters in 1973, the Peters world map is still controversial. Although it has many passionate advocates, more traditionalist cartographers, who hold it to be more of a political statement than a cartographic one, have fiercely attacked it. They say that, contrary to his claims at the time, the projection Peters used was by no means original and that, rather than being area accurate, the map is distorted, being crushed toward the poles and stretched across the equator. Peters himself dismissed all existing world maps as an expression of "the age in which the white man ruled the world." The map's presentation of the developing world as larger than conventionally shown impressed itself on the public imagination in an age of increasing concern about world inequalities. The projection was adopted by several UN agencies and favored by developing countries, who felt that the traditional Mercator projection exaggerated the developed world.

instance, was shown as being only 18 percent larger than it was as opposed to the 223 percent for maps based on the Van der Gritten projection. For its part, the USA shrank by 3 percent.

Another more recent and potentially extremely influential projection is the Hobo-Dyer projection (see p.180), created by British cartographer Mick Dyer by modifying the 1910 Behrmann projection in 2002. Like the Peters projection, maps created using it are equal size with landmasses shown in true proportion. However, it is far more accurate than the Peters in shape. The projection is double-sided, each side with its own individual worldview. One view is Africa-centered with the North Pole at the top, while the other is centered on the Pacific with the South Pole at the top.

CARTOGRAPHY REVOLUTIONIZED

Isometric maps are just one of the new forms of mapping that emerged as a result of a cartographic revolution that began in around the early 1980s. This revolution was to affect the nature of maps—what they were used to show—the way in which the data required to compile them were collected, processed, and manipulated; and how maps were produced and disseminated.

The prime technological drivers in this truly radical transformation were the advent of the personal computer and, subsequently, of the Internet; and the use of space satellites for data collection, and the consequent development of Global Positioning Systems (GPS) and what are known as Geographical Information Systems (GIS).

Global Positioning Systems

Without modern satellite technology, GPS would have been impossible to conceive, let alone develop and implement. Put simply, it is an umbrella term, used to describe the way in which the Earth can be measured from space and the positions of places and features on it recorded. From around 1987 onward, GPS started replacing traditional surveying methods and superseding the ways in which latitude, longitude, and elevation had hitherto been determined.

The transformation began in 1972, when NASA put the first remote-sensing satellite into orbit and so inaugurated the era of satellite mapping. This was the first of many such launches. Today, a global network of 24 orbiting satellites—and receiving stations that translate the signals from the satellites into readable positional information—can provide surveyors and mapmakers with geographical co-ordinates for any surface feature on the planet in minutes, as well as real-time computer-generated terrain images. During the Gulf War of 1991–2, the system, which until then had been restricted to military use, became accessible to all users. It can now be utilized by anyone with a GPS receiver. The similar Russian system is called GLONASS, and the European Union system, which is planned to become operational in 2008, is named Galileo. The data such systems produce can be utilized to create all kinds of maps, including ones devoted to topics such as environmental change, agriculture and other forms of land use, and population growth, density, and distribution.

Landsat mapping

Like the GPS, the Landsat program got under way in 1972, when the USA launched its first remote-sensing satellite from what was then Cape Kennedy and is now once again Cape Canaveral. The satellite was designed to record mappable data with a multi-spectral scanner that built up digital images by measuring waves of light and heat energy emitted by or reflected from the planet's surface. The results were then transmitted back to receiving stations on Earth. The technique is known as remote sensing.

Early GPS equipment, Prakla–Seismos, 1978
GPS is a system of satellites in Earth orbit that broadcasts signals, allowing ground-based equipment to calculate its position and elevation. Modern GPS receivers are small enough to be carried by hand.

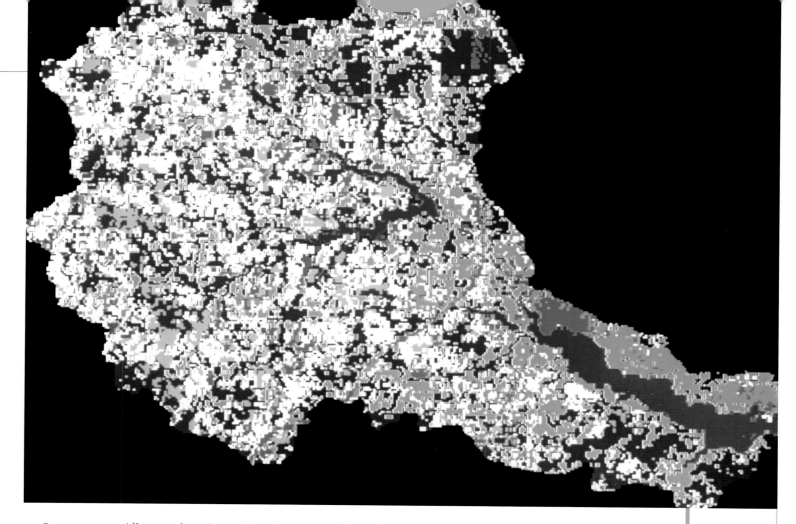

By measuring different infrared wavelengths, it is possible to focus on different aspects of the planetary surface. As satellite technology advanced, the possibilities naturally expanded into the bargain. The first Landsat satellite, for instance, had only four channels in its multi-spectral scanner. By the time Landsat 4 was launched ten years later—followed by Landsat 5 in 1984—the number of channels in the satellites' newly developed enhanced thematic mappers had been increased to seven. This made it possible to penetrate further into the infrared wavelengths, revealing colors that are invisible to the naked eye. Landsat 7, launched in 1999, was even more advanced. It is one of the two Landsat satellites that are still operational, the other being Landsat 5. Both satellites continue to provide coverage of the entire globe every 16 days as they scan 113-mile (182-km) bands of the planet from an orbital height of 437 miles (703 km).

The program has greatly enhanced human understanding of the geography of the planet, particularly changes in land use. Landsat satellites have proved to be the key to providing more solid information on rates of deforestation, desertification, and urban growth across the world. The systems they use are a natural development of early low-resolution ones, which, though less sophisticated in comparison, were still good enough to plot weather systems. It is increasingly improved resolution that has allowed much more accurate mapping. The equivalent French and Russian programs are equally sophisticated.

Seasat and Geosat

The Landsat satellites are not the only ones to be actively engaged in mapping the Earth from space. Other satellites are dedicated to observing planetary weather patterns and other aspects of meteorology and the world's oceans and seas. Seasat (see p.178), launched in 1978, had only a brief career, but Geosat, launched by the US Navy in 1985, has had a much longer one. It was designed to measure the height of the seafloor by bouncing a radar beam off it. The result was the most comprehensive set of gravity measurements ever. When the data was declassified in 1995,

97 Satellite-Derived Land-Use Map, Landsat 5, NASA, USA, 1987

Covering the Navesink watershed in the northeastern part of the state of New Jersey, this map shows how data provided by the Landsat satellite can be utilized by cartographers to create all kinds of maps: in this case, to assist in land planning by helping to determine what effects agriculture and urban activity have on water pollution in the area. The red represents high-density urban areas and the purple medium- to low-density ones. Blue and aqua respectively denote water and wetlands, while yellow is cropland, green is forest, and white is pasture or vacant or barren land.

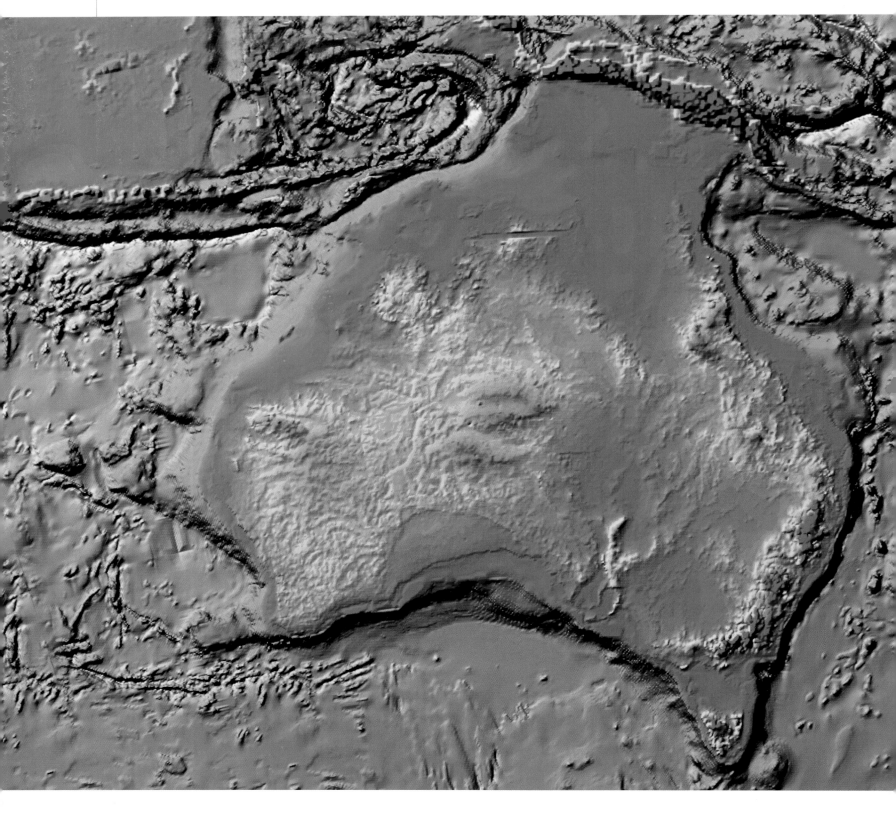

American marine geophysicists David Sandwell and Walter Smith wasted no time in using it to create their gravity map—the best view to date of what the ocean would look like if it were waterless.

Mapping the environment

In 1999, the USA launched Terra. This is an orbital satellite the size of a small school bus equipped with five state-of-the-art scanners, which collect nearly 20 terabytes of data every three months. It is one of a series of satellite sensors that have been placed in orbit by NASA's Earth Observing System (EOS).

98 Sea and Land Topography of Australia, Seasat, NASA, USA, 1978

This map was constructed from radar data transmitted by the Seasat space satellite, the first satellite to be specifically tasked with the study of the oceans. The colors represent altitude, from gray—the deep ocean— through blue, green, and yellow to red for the map's highest points. The boundary between land and sea is in the mid-green tones, while strong relief features are shaded. Seasat ceased to function just less than four months after its launch in 1978 as a result of a massive short circuit in its electrical system.

The data Terra collects as it orbits from pole to pole is related to the Earth's atmosphere, land, oceans, and radiant heat. The high-resolution images transmitted by ASTER (Advanced Spaceborne Thermal Emission and Reflection Radiometer) have been used to create elaborately detailed three-dimensional maps of planetary landforms. CERES (Clouds and Earth's Radiant Energy System) maps radiant heat patterns, and MISR (Multi-angle Imaging Spectro-Radiometer) monitors climatic changes. MODIS (Moderate-resolution Imaging Spectro-Radiometer) maps cloud cover, ice and snow extent, and the distribution of phytoplankton, the minute plants that live in the world's oceans. MOPIT (Measurement of Pollution in the Troposphere) tracks and monitors carbon dioxide and methane emissions from urban and industrial areas.

In the late 20th century, increasing global interest in the environment led to a rise in the number of environmental maps produced. As concern about the potentially irreversible damage unchecked human activity is inflicting on the natural environment has heightened, so mapmakers have responded by producing maps that chart the effects of pollution, the loss of biodiversity, and other forms of environmental degradation, such as the loss of the ozone layer and the potentially disastrous effects of global warming. On the whole, mapmakers simply map what scientists and environmentalists have worked out, to illustrate situations that they believe to exist. However, satellite images over a period of years have allowed the charting of the loss of rainforests, such as the ones in Amazonia and Indonesia.

The European Space Agency's Envisat environmental satellite is currently engaged in the creation of the most detailed cartographic portrayal of the Earth's land surface in human history. The Globcover project, as it is termed, aims to produce a unique depiction of the face of the planet, broken down into more than 20 separate land-use classes or categories. The latest estimate is that up to 20 terabytes of imagery will be required to create the final Globcover map, an amount of data equivalent to the contents of 20 million books. The finished map will prove to be a vital tool for the monitoring of the health of the planet and the changing environment.

Extraterrestrial mapping

Extraterrestrial mapping, starting with the lunar mapping programs of the USA and the then USSR and continuing with the American mapping of Mars, Venus, and the moons of Jupiter and Saturn, has taken cartography beyond the Earth and out into the far reaches of outer space.

In 1960, the USSR produced the first map of the far side of the Moon; later, the USA's Lunar Orbiter and Apollo Moon missions produced photographs that enabled cartographers to produce topographic maps of much of the Moon's surface. Between 1990 and 1995, Magellan space probes used imaging radar to map almost the entirety of Venus. The Mars Global Surveyor probe of 1997, data from which was employed to produce the first three-dimensional map of the so-called red planet, followed.

Such cartography could be argued to have dramatically changed the late 20th-century worldview, by altering for ever our conception of the boundaries of our "world." This now consists of the Earth plus the Moon. Skylab photographs and views from the Moon of the Earth have certainly put the latter in perspective.

NEW MAPS FOR NEW INFORMATION

GIS is the blanket expression used to describe the technology that makes use of computers, mapping software, and digitized geo-referenced data to generate maps electronically. Thematic mapping took off in the 19th century, when maps became useful tools for analysis and planning. The ease with which such maps are now made depends on linking spatial statistical analysis and mapping software.

Demography

The mapping of demography was a typical example. Although such cartography was not dependent on GIS—early versions were made even without computers—computer assistance made it much easier to carry out such mapping. Before this, such maps had often been unable to present a complete picture: like all maps, they could be, by definition, only as good as the information used as the base for their compilation.

Back in 1870, for instance, Francis Amasa Walker (1840–97) produced what is generally regarded as a landmark *Statistical Atlas of the United States*. The atlas contained a total of 54 maps covering the physical, economic, and social geography of the USA. Through them, among other things, Walker attempted to map the locations of different racial and ethnic groups and chart the social differences between them. The atlas set a precedent as far as its use of census information was concerned, but the majority of its social data maps were confined to the area of the country east of the 99th meridian. By comparison, the coverage of the Pacific area was scanty and incomplete.

Since it relies so heavily on the ready availability of large amounts of reliable data, and particularly on census returns, demographics as a science really came into its own only in the 20th century. Even today, particularly in the developing world, such data are still difficult to obtain, since many people equate census time with subsequent attempts to impose raised taxation and they leave the area.

99 **The Population Map, ODT Inc., USA, 2005** This population cartogram illustrates demography at work on a global scale. The idea behind the map is to relate the size of countries to their respective populations, with each square on the grid representing a million people. Visually, this makes China, rather than Russia, the biggest country on the planet, with India not far behind. In contrast, the USA has dramatically shrunk in size—it is far smaller than Indonesia—because it contains only 4.5 percent of the world's population, while nations with populations under a million do not make it onto the cartogram at all. The map was conceived by Bob Abramms, ODT's founder, while cartographer Paul Breeding was responsible for the main cartogram. His achievement was to devise a means of presenting population statistics accurately while retaining as much of the geographical shape of the countries as possible. The inset map is the equal-area Hobo-Dyer world map Abramms and his associate Howard Bronstein commissioned from British cartographer Mick Dyer in 2002. As its name suggests, it shows the land area of countries in true proportion.

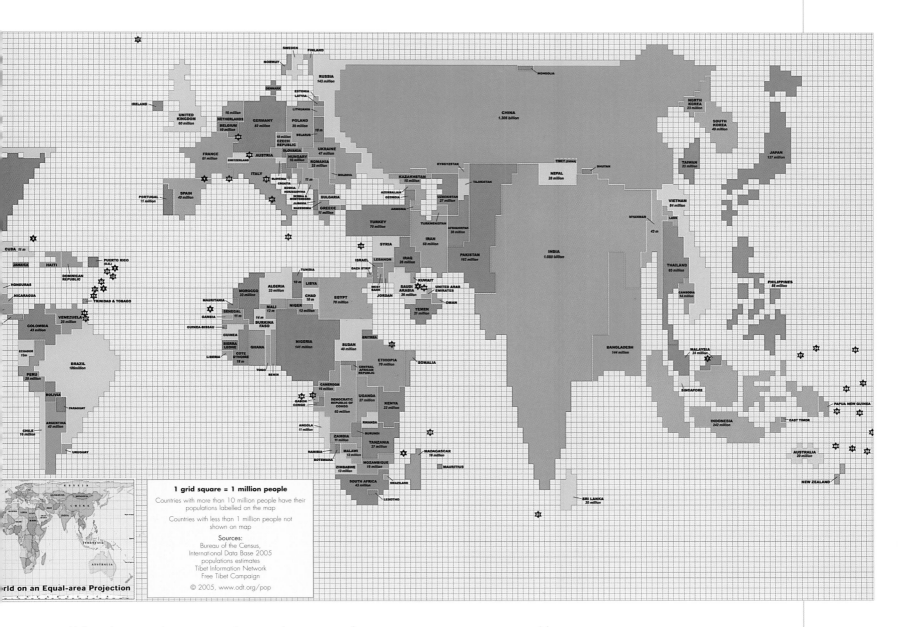

1 grid square = 1 million people

Countries with more than 10 million people have their populations labelled on the map

Countries with less than 1 million people not shown on map

Sources:
Bureau of the Census,
International Data Base 2005
populations estimates
Tibet Information Network
Free Tibet Campaign

© 2005, www.odt.org/pop

...rld on an Equal-area Projection

What demography is now indicating, however, is that, contrary to expectations, world population growth is actually slowing and may well eventually come to a virtual standstill. It is now expected to peak at 9 billion by 2070. This contradicts previously held theories, which in part can be traced back to the great British economist Thomas Malthus (1766–1834), who proposed that, rather than the proliferation of weapons, the major threat to global stability was uncontrolled population growth and a consequent outstripping of planetary resources.

Measuring poverty

In the 20th century, cartography was increasingly used to alert the public to the reality or possibility of human catastrophes, such as maps showing the occurrence of famine, or the dramatic map of sub-Saharan Africa produced for the Central Intelligence Agency in 2000 to highlight the prevalence of HIV in the region. The Human Development Index was originally conceived by Mahbub ul-Haq (1934–88), a one-time Director of the World Bank. It was adopted by the United Nations in 1993

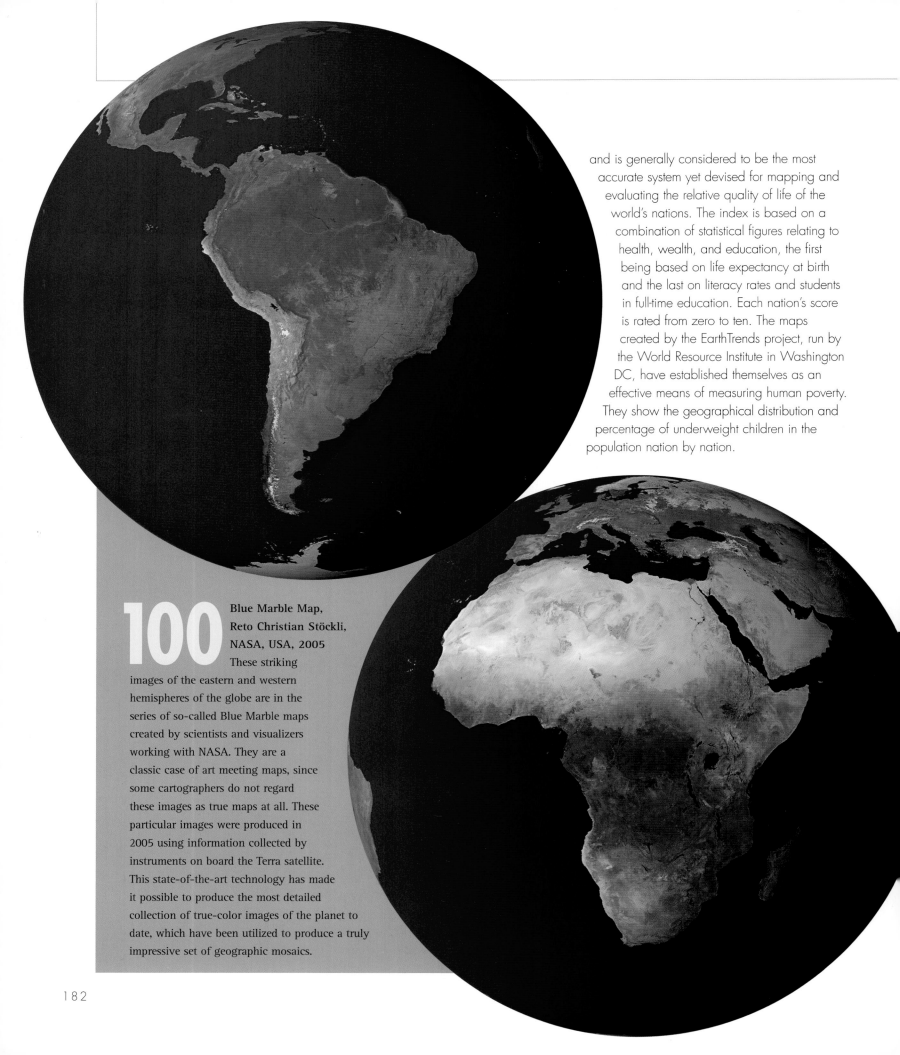

and is generally considered to be the most accurate system yet devised for mapping and evaluating the relative quality of life of the world's nations. The index is based on a combination of statistical figures relating to health, wealth, and education, the first being based on life expectancy at birth and the last on literacy rates and students in full-time education. Each nation's score is rated from zero to ten. The maps created by the EarthTrends project, run by the World Resource Institute in Washington DC, have established themselves as an effective means of measuring human poverty. They show the geographical distribution and percentage of underweight children in the population nation by nation.

100 Blue Marble Map, Reto Christian Stöckli, NASA, USA, 2005 These striking images of the eastern and western hemispheres of the globe are in the series of so-called Blue Marble maps created by scientists and visualizers working with NASA. They are a classic case of art meeting maps, since some cartographers do not regard these images as true maps at all. These particular images were produced in 2005 using information collected by instruments on board the Terra satellite. This state-of-the-art technology has made it possible to produce the most detailed collection of true-color images of the planet to date, which have been utilized to produce a truly impressive set of geographic mosaics.

EVERYONE'S A MAPMAKER

As humanity takes its first steps into the new millennium, the question of what kind of maps will be made over the coming years and how they will be produced remains open. The chief point at issue is whether the traditional printed paper map still has a part to play in this future, or whether, as some argue, it may be replaced over time by the electronic products of computer and digital technology. The same argument has been applied to books themselves.

Virtual maps, available on the Internet or as software packages, are rapidly becoming a major map form of the early 21st century. It has been calculated that, in 2002, nearly half of all Internet users in the USA—that is, approximately 87 million people—consulted a virtual map at least once a month, while, as far back as 1996, a single Californian website was processing over 90,000 map requests a day. The number of sites featuring virtual maps—both static and interactive—runs into the tens of thousands; where appropriate, many of these maps, such as ones showing weather patterns or traffic conditions, are updated throughout the day.

One of the chief advantages of virtual maps is that they can, in theory, be made fully interactive. Depending on the software—the availability of which is still currently limited—elements of such a map can be enlarged or reduced in size, or the map as a whole can be rescaled at the touch of a computer key. Colors, typestyles, and symbols can be altered just as easily. The information can be presented by choosing to highlight layers of information either separately or together. The electronic linking of such maps to other maps means that the amount of information that can be accessed is almost limitless. The availability of downloadable map data also means that today's mapmakers are spoilt for choice. Put at its simplest, provided that they possess the necessary software and skills, everyone who wants to can be a mapmaker.

However, it is likely that printed maps will be with us for many years to come. Virtual maps are just that: unless they are saved or printed out, they vanish from the screen as soon as the computer is turned off. In the main, most computer screens still cannot match the resolution, color quality, and size achieved by the printed page. One of the major potential weaknesses of maps imported into and exported out of the virtual world is their provenance—and that of the information on which they are based. If this is unreliable, the map, however attractive it may appear, is of questionable value. The same, of course, could be said for paper maps, but today those have usually been authenticated before publication.

The next generation

It is likely that—as both the map-using public and map-making technology become increasingly sophisticated—the near future will see a creative marriage between traditional and digital cartography, with the creation of new, visually stunning, ultra-informative maps. The "Blue Marble" maps, produced by Reto Christian Stöckli, a Swiss-based scientist who is NASA's scientific visualizer, and his colleagues from 2000 onward, can be seen as part of this new generation of maps, although some mapmakers argue that these images are very sophisticated photographs and not true maps. To become maps, they say that they must at least show the equator, label some features, and give an explanation of the colors used. But this does not alter the fact that the Blue Marble maps are extremely beautiful, with the Earth portrayed in true color and as though viewed from a distance of 22,000 miles (32,000 km).

Plans for the next generation of these maps are already being made. According to Stöckli, the idea is to produce them on a month-by-month basis to illustrate the Earth's climatic cycle. After that, who knows what the next step will be, but the same might well be said of cartography as a whole.

CHRONOLOGY

c. 13000 B.C.E. The Yunta rock engravings are made, in Australia.

c. 10000 B.C.E. One of the oldest rock maps in North America is created in Idaho.

c. 6200 B.C.E. Mapmakers at Çatal Hüyük, Turkey, create the world's earliest known surviving town plan.

c. 2300 B.C.E. Mesopotamians map the city of Lagash.

c. 2200 B.C.E. Babylonians incise the earliest known cadastral maps onto clay tablets.

c. 1500 B.C.E. Plans of Nippur, the Sumerian capital just south of Babylon, are created.

c. 1300 B.C.E. The Turin Papyrus, the earliest known topographical map from ancient Egypt, is compiled.

c. 1200 B.C.E. Rock engravings are made at Val Camonica, near Brescia in northern Italy, followed by others dating from c. 800 B.C.E.

c. 600 B.C.E. The Babylonian world map is created.

c. 600 B.C.E. Anaximander of Thales is believed to have produced a world map, no details of which survive.

c. 200 B.C.E. Chinese geographers conduct a cartographical survey of Southeast Asia.

c. 134 B.C.E. Hipparchus of Rhodes compiles the earliest known star chart.

c. 7 B.C.E. Marcus Vipsanius Agrippa's *Orbis Terrarum* (Survey of the World) is completed.

c. 150 Claudius Ptolemaeus (Ptolemy) writes his *Geographike Hyphegesis* (Geography).

c. 203–208 The Romans create the Forma Urbis Romae, a monumental carved street plan of the imperial capital.

276 The Chinese scholar and geographer Pei Xian writes his *Six Laws of Mapmaking*.

4th century The Peutinger Table, a route map of the Roman Empire, is compiled. It survives as a 13th-century copy.

c. 350 The *Corpus Agrimensarum*, a collection of the most important Roman land surveys, is compiled.

c. 400 The Roman thinker Ambrosius Aurelius Theodosius Macrobius puts forward the notion that the world is divided into zones according to climate.

c. 550 The Madaba mosaic, the largest and most detailed map to survive from Byzantine times, is created in Palestine.

c. 776 A Spanish monk, Beatus of Liébana, creates his *mappa mundi* (world map).

10th or early 11th century The Anglo-Saxon *mappa mundi* is drawn. It is thought most likely to be a debased copy of a Roman original.

1154 Islamic scholar Al-Idrisi's study of world geography, the *Book of Roger*, is completed in Sicily.

1155 A map of western China, which may be the world's first printed map, is created by Chinese geographers.

c. 1250 Scholar-monk Matthew Paris maps England.

c. 1290 The Carte Pisane, the oldest surviving portolan (navigational) chart, is compiled in Italy.

c. 1290–1300 The Hereford *mappa mundi* is produced.

c. 1300 The Ebstorf *mappa mundi* is compiled in Germany.

c. 1360 The Gough map, the first detailed route map of Britain, is created.

1375 Abraham Cresques produces his Catalan Atlas.

1472 The first Western printed map is produced in Germany.

1475 Having been translated into Latin in Byzantium in 1406, Ptolemy's *Geography* is printed in Italy.

c. 1480 German mapmaker Hans Rüst's world map is the first to be printed in the vernacular, rather than Latin.

c. 1492 Martin Behaim completes his globe, the oldest known globe to survive.

c. 1500 Erhard Etzlaub's Rom Weg (The Way to Rome) is the first route map to be printed.

1502 The Cantino planisphere charts Portuguese and Spanish discoveries in the New World.

c. 1502 Leonardo da Vinci draws a plan of Imola, in Italy. It is one of the earliest Renaissance geometric town views to be made.

1507 Martin Waldseemüller's world map is the first to name "America."

1513 The Ottoman sailor and geographer Piri Reis creates a world map that demonstrates that knowledge of Columbus's discoveries has spread to the Arab world.

1533 Dutch mapmaker Gemma Frisius demonstrates how to determine map locations by triangulation.

1537 The great Flemish mapmaker Gerardus Mercator compiles his first map.

1542 Jean Rotz presents his *Boke of Idrography* (Book of Hydrography) to Henry VIII of England.

1569 Mercator's world map employs his newly devised projection.

1570 Abraham Ortelius publishes the *Theatrum Orbis Terrarum* (Theater of the Earth), the first true atlas.

1572 Georg Braun and Franz Hogenberg publish the first volume of their *Civitates Orbis Terrarum* (Cities of the World), the start of a multi-volume compilation of city maps.

1578 Christopher Saxton completes his pioneering survey of England and Wales.

1584 Lucas Janszoon Waghenaer publishes *Spiegel der Zeevaert* (Mirror of the Sea), the first sea atlas to be printed.

1585 Mercator, the first to coin the term "atlas," publishes the first edition of his world atlas.

c. 1587 The Englishman John White surveys the Virginian coast.

1609 Frenchman Samuel de Champlain maps the eastern coast of North America from Cape Cod to Cape Sable.

1656 Frenchman Nicholas Sanson maps Canada.

1662 Joan Blaeu publishes his *Atlas Major* in the Netherlands.

1669 Jean Picard measures a degree of latitude on the Paris meridian.

1675 John Ogilby completes the first English road atlas.

1701 Edmond Halley pioneers the first contour maps.

1718 Frenchman Guillaume Delisle maps Louisiana and the course of the Mississippi.

1733 Henry Popple publishes his map of the North American colonies, America Septetrionalis.

1744 Jacques Cassini completes his map of France. It is the first survey of an entire country to be based on scientific triangulation.

1755 John Mitchell publishes his map of English and French landholdings in North America.

1761 John Harrison devises a "marine timekeeper." Its accuracy means that longitude can finally be calculated accurately.

1768 Charles Mason and Jeremiah Dixon survey the boundary between Pennsylvania and Maryland. The so-called Mason-Dixon line comes to symbolize the division between the non-slave states of the North and the slave states of the South.

1769 James Cook charts the coasts of New Zealand and parts of Australia.

1782 James Rennell, the first Surveyor-General of Bengal, completes his map of "Hindoostan" for the East India Company. It is published in 1788.

1785 In the USA, the Continental Congress enacts the Land Ordnance, dividing the new nation into townships and ranges.

1791 The Trigonometrical Survey of Britain commences. It is renamed the Ordnance Survey in 1801.

1798 In the USA, Valentine Seaman produces the earliest known maps to show the incidence of disease.

1802 Work starts on the Great Trigonometrical Survey of India.

1804–06 Meriwether Lewis and William Clark explore and map America west of the Mississippi.

1811 The naturalist and explorer Alexander von Humboldt publishes his *Atlas Geographique et Physique*.

1815 William Smith publishes his geological survey of England and Wales.

1826 Pierre Dupin produces a cartogram of France showing rates of illiteracy.

1838 The US Corps of Topographical Engineers is established.

1842 John Charles Freemont leads the first of three major surveying expeditions to the Rocky Mountains, Oregon, California, and Upper California.

1845 German geographer Heinrich Berghaus produces his *Physikalischer Atlas*, one of the first thematic atlases to be compiled.

1861 Francis Galton produces the first modern weather map.

1867–79 The four greatest surveys of the American West are carried out, by Clarence King, Ferdinand Hayden, John Wesley Powell, and George Wheeler.

1870 Francis Amasa Walker produces his *Statistical Atlas of the United States*.

1879 The US Geological Survey is established.

1889 Charles Booth produces his Descriptive Map of London Poverty.

1921 In the USA, the Schoolcraft Quadrangle is surveyed from the air.

1933 Harry Beck's schematic map of the London Underground, a renowned design classic, is published.

1936 Phyllis Pearsall's London A–Z street atlas is produced.

1950 Orthophoto mapping, using modified aerial photographs, is developed.

1957 Marie Tharp and Bruce Heezen publish their Physiographic Map of the North Atlantic, a groundbreaking map of the seafloor.

1960 The first meteorological satellite is launched.

1960 The USSR produces the first map of the far side of the Moon.

1968 Apollo 8 takes the first pictures of the Earth from space.

1972 The first Landsat satellite is launched, allowing the mapping of land-use.

1973 German historian and cartographer Dr. Arno Peters puts forward his revolutionary equal-area world map projection.

1978 The Seasat satellite is launched, followed by Geosat in 1985.

1981 Marie Tharp, Alvaro Espinosa, and Wilbur Rinehart compile their World Earthquake Map.

1987 Global Positioning System (GPS) starts to replace traditional surveying methods.

1989 The first CD-Rom map of the world is produced.

1993 The first interactive map appears on the Internet.

1993 The UN adopts the Human Development Index for mapping the relative quality of life in the world's nations.

1999 Terra satellite is launched to enable environmental conditions on the Earth to be mapped more thoroughly.

2000 The *Endeavour* space shuttle maps 47.6 million square miles (123 million sq km) of Earth from space.

FURTHER READING

Books

Bagrow, Leo, *History of Cartography*, ed. R.A. Skelton, 2nd revised edition, Precedent, 1985.

Barber, Peter (ed.), *The Map Book*, Weidenfeld Nicolson Illustrated, 2005.

Barber, Peter and Board, Christopher (eds.), *Tales from the Map Room*, BBC Books, 1993.

Bendall, Sarah (ed.), *Dictionary of Land Surveyors and Mapmakers of Great Britain and Ireland 1530–1850*, British Library, 1997.

Bendall, Sarah, *Maps, Land, and Society: A History*, Cambridge University Press, 1992.

Black, Jeremy, *Maps and History*, Yale University Press, 1997.

Blake, John, *The Sea Chart*, Conway Maritime, 2004.

Brown, Lloyd A., *The Story of Maps*, Dover Publications, 1980.

Brotton, Jerry, *Trading Territories: Mapping the Early Modern World*, Reaktion Books Ltd., 1997.

Buisseret, David, *The Mapmakers' Quest*, Oxford University Press, 2003.

Buisseret, David (ed.), *Monarchs, Maps and Ministers*, University of Chicago Press, 1992.

Buisseret, David (ed.), *Rural Images: Estate Maps in the Old and New Worlds*, University of Chicago Press, 1996.

Campbell, Tony, *The Earliest Printed Maps 1472–1500*, British Library and University of California Press, 1987.

Campbell, Tony, *Early Maps*, Abbeville, 1981.

Crone, Gerald R, *Maps and Their Makers*, 5th edition, Dawson UK, 1978.

Delano-Smith, Catherine and Kain, Roger J.P., *English Maps: A History*, British Library, 1999.

Dilke, O.A.W., *Greek and Roman Maps*, Thames & Hudson, 1985.

Edney, Matthew H., *Mapping an Empire: The Geographical Construction of British India 1765–1843*, University of Chicago Press, 1990.

Ehrenberg, Ralph E., *Mapping the World*, National Geographic Books, 2005.

Goss, John, *The Mapmaker's Art: A History of Cartography*, Studio Editions, 1993.

Harley, J.B. and Woodward, D., *The History of Cartography*, University of Chicago Press. Vol. 1, *Cartography in Prehistoric, Ancient and Medieval Europe and the Mediterranean*, 1987, Vol. 2 Book 1, *Cartography in the Traditional Islamic and South Asian Societies*, 1992, Vol. 2 Book 2, *Cartography in the Traditional East and Southeast Asian Societies*, 1994, Vol. 2 Book 3, *Cartography in the Traditional African, American, Arctic, Australian and Pacific Societies*, 1998, Vol. 3, *Cartography in the European Renaissance*, in press, Vols. 4–6 in preparation.

Harris, Nathaniel, *Mapping the World: Maps and their History*, Brown Partwork and Thunder Bay Press, 2002.

Harvey, P.D.A., *The History of Topographical Maps*, Thames & Hudson, 1980.

Harvey, P.D.A., *Mappa Mundi: The Hereford World Map*, British Library, 1996.

Harvey, P.D.A., *Maps in Tudor England*, The British Library/Public Record Office, 1993.

Harvey, P.D.A., *Medieval Maps*, British Library, 1991.

Hodgkiss, Alan G., *Discovering Antique Maps*, Shire Publications, 1996.

Kain, Roger J.P., Chapman, John and Oliver, Richard R., *The Enclosure Maps of England and Wales*, Cambridge University Press, 2004.

Keay, John, *The Great Arc: The Dramatic Tale of How India was Mapped and Everest was Named*, HarperCollins, 2000.

Konvitz, Josef W., *Cartography in France 1660–1848*, University of Chicago Press, 1987.

Kretschmer, Ingrid, Dörflinger, Johannes and Wawrik, Franz (eds.), *Lexikon zur Geschichte der Kartographie*, 2 Vols., Deuticke, 1986.

Lynam, Edward, *The Mapmaker's Art*, Batchworth, 1953.

Monmonier, Mark, *How to Lie with Maps*, University of Chicago Press, 1991.

Monmonier, Mark, *Rhumb Lines and Map Wars*, University of Chicago Press, 2004.

Nebenzahl, Kenneth, *Maps from the Age of Discovery*, Random House, 1990.

Norwich, Oscar I., *Maps of Africa*, Ad. Donker, 1983.

Robinson, Arthur H., *Early Thematic Mapping in the History of Cartography*, University of Chicago Press, 1982.

Schilder, Günter, *Monumenta Cartographica Neerlandica*, 7 Vols., Uitgeverij Canaletto/Repro Holland BV, 1986–2003, Vols. 8–9 in preparation.

Shirley, Rodney W., *The Mapping of the World: Early Printed World Maps 1472–1700*, 4th edition, Early World Press, 2001.

Skelton, R.A., *Decorative Printed Maps of the 15th to 18th Centuries*, Spring Books, 1952.

Short, John Rennie, *Representing the Republic*, Reaktion Books, 2001.

Snyder, John P., *Flattening the Earth: Two Thousand Years of Map Projections*, University of Chicago Press, 1993.

Suarez, Thomas, *Early Mapping of the Pacific*, Periplus Editions, 2004.

Tooley, Ronald V., Bricker, Charles and Crone, Gerald R., *Landmarks of Mapmaking: An Illustrated Survey of Maps and Mapmakers*, Phaidon, 1976.

Wallis, Helen M. and Robinson, Arthur H. (eds.), *Cartographical Innovations: An International Handbook of Mapping Terms to 1900*, Map Collector Publications in association with the International Cartographic Association, 1987.

Whitfield, Peter, *The Charting of the Oceans*, Pomegranate Communications, 1996.

Whitfield, Peter, *The Image of the World*, British Library, 1994.

Whitfield, Peter, *New Found Lands*, British Library, 1998.

Wilford, John Noble, *The Mapmakers*, Knopf, 2000.

Wolter, John Amadeus and Grim, Ronald E. (eds.), *Images of the World: The Atlas Through History*, McGraw-Hill, 1997.

Wood, Denis, *The Power of Maps*, The Guilford Press, 1992.

Woodward, David (ed.), *Five Centuries of Map Printing*, University of Chicago Press, 1975.

Journals

Imago Mundi: The International Journal for the History of Cartography
A scholarly periodical devoted to the history of maps.
www.maphistory.info/imago.html (journal content)
www.tandf.co.uk/journals/titles/03085694.asp (subscriptions and sales)

Journal of Cartography and Geographic Information Science
The official journal of the International Cartographic Association (ICA).
www.cartogis.org/publications/cagisjournal

Cartographica
The international journal for geographic information and geovisualization, with articles on the history of cartography.
www.utpjournals.com/jour.ihtml?lp=carto/carto.html

MapForum
A specialist antique map magazine.
www.mapforum.com

Websites

oddens.geog.uu.nl/index.php
Oddens' bookmarks, with a list of map links.

www.bl.uk/collections/map_overview_history.html
The British Library Map Room is home to the second largest map collection in the world, after the Library of Congress, with over 4.25 million atlases and maps.

www.lib.utexas.edu/maps
The website at the University of Texas houses a vast historical map collection, including maps originated by the CIA.

www.lib.virginia.edu/exhibits/lewis_clark
The University of Virginia website is mainly concerned with exploration, with a particular focus on the Lewis and Clark expedition.

www.loc.gov/rr/geogmap/
The United States Library of Congress has the largest and most comprehensive cartographic collection in the world, with collections numbering over 5 million maps and 72,000 atlases.

www.maphistory.info/mapsindex.html
A gateway site and the best starting-point for researching any particular line of interest, with a good index and extensive links.

www.nationalgeographic.com/maps
Fascinating examples of the latest interactive maps.

www.newberry.org/smith/smithhome
A website run by the Hermon Dunlap Center for the History of Cartography at the Newberry Library in Chicago. This contains a fine collection of maps, with some interesting essays and commentaries.

INDEX

ACKNOWLEDGMENTS

Sources: AA= The Art Archive, AI = akg-images, BAL = The Bridgeman Art Library, BL = The British Library, LPC = The Lordprice Collection

Front cover: Detail from Gerardus Mercator's *Atlas* 1619, BL/BAL.
Back cover: Corbis/Gianni Dagli Orti (top); AI (bottom).

Pages: 1 LPC/Private Collection; 2-3 AI; 6 AI; 7 BL, London; 8 Science & Society Picture Library; 9 Werner Forman Archives; 10-11 The British Museum; 12 Corbis; 13 Cambridge University/James Mellaart; 14 Wayne T. Crans/www.idahorockart.com; 17 AA/Dagli Orti; 18 AA/Egyptian Museum, Turin/Dagli Orti; 19 Corbis; 20-21 BAL/Musee de la Poste, Paris; 22 BL, London Harley 3686 f.36v; 23 Marshall Editions, London; 24-25 AA/Dagli Orti; 26 Stanford's Digital Forma Urbis Romae Project/Stanford University; 28 AA/Dagli Orti; 29 The Library of Congress; 30-31 Corbis/courtesy of Museum of Maritimo, Barcelona/Ramon Manent; 32 BAL/Municipal Library, Porto, Portugal; 33 BL, London/Add. 22797 f.99v; 34-35 BAL/Private Collection; 36 BAL/BL, London/Cott Tib B V Part 1 f.56v; 38 AI; 41 BAL/The British Museum; 42 BAL; 43 BL, London/Ms Harl 2772.fik.70v; 44 University Library Leiden/Collection Bodel Nijenhuis/MS Or3101 fols. 4-5; 45 AI/Jean-Louis Nou; 46 AA/Bodleian Library Oxford, Pococke 375 folio; 48 Bibliotheque Nationale, Paris; 49 BAL/The National Maritime Museum; 50 AI/BL, London; 51 AI/BL, London; 52 Bodleian Library, Oxford; 53 The National Archives/Public Records Office, Kew; 54 Corbis/Gianni Dagli Orti; 55 BL, London; 57 Scala, Florence/Marciana Library, Venice, Italy; 58 LPC/Private Collection;

60 AI/Visioars; 61 Royal Library, Copenhagen; 62 Scala, Florence/Biblioteca Estense, Modena; 65 AI; 66 The Library of Congress; 67 AI; 69 www.prep.mcneese. edu/courtesy of Dr. Joseph Richardson; 70 AA/Museo Ciudad Mexico/Dagli Orti; 71 BAL; 72 Bibliotheque Nationale, Paris; 73 Corbis/Gianni Dagli Orti; 74 AA; 77 BL, London; 78 The Library of Congress; 79 AA/BL, London; 80 Corbis/The Bettmann Archive; 82-83 BAL/BL, London; 84-85 BL, London; 86 BAL; 87 BL, London; 88-89 LPC/Private Collection; 90-91 LPC/Private Collection; 93 Scala, Florence/Maps Gallery, Vatican; 95 BAL/Institut Tessin, Hotel de Marle, Paris; 96 LPC/Private Collection; 99 LPC/Private Collection; 100 Scala, Florence/Museo Correr, Venice; 101 Scala, Florence/Museo Vinciano, Vinci; 102 BAL/Private Collection; 103 LPC/Private Collection; 104 The National Maritime Museum; 106 Corbis; 108 AI/Musee de l'Armee, Paris; 109 BL, London/Add or Sloane ms4723; 110 BL, London; 112 Scala, Florence/New York Public Library; 113 Corbis/The Bettmann Archive; 114 LPC; 115 The National Archives/Public Records Office, Kew; 116 BAL; 117 Scala, Florence/HIP/The British Museum; 118 AI; 119 The National Maritime Museum; 120 Art Gallery @ New South Wales © Dr. David Malangi/DACs; 122 BAL; 123 The National Maritime Museum; 124 The Royal Geographical Society; 125 Cambridge University Library; 126-127 AA/Private Collection/Laurie Platt Winfrey; 128 BAL/Private Collection; 130-131 AA/Bodleian

Library, Oxford/by Capt J.C.R. Colomb. N.12288b after p 88; 132 LPC; 133 The National Archives/Public Records Office, Kew; 135 BL, London; 136-137 The Library of Congress/Geography & Map Division, Washington DC; 138 Corbis/The Bettman Archive; 139 National Library of Scotland, NLS shelfmark: EMS.s.156 reproduced by permission of the Trustees of the National Library of Scotland; 140 BAL; 141 Virginia Historical Society, Robert Snedden Mss5:1 v.3; 142 LPC; 145 LPC; 146-147 LPC; 148 BAL/Collection Kharbine-Tapabor, Paris, France; 149 Corbis/The Bettmann Archive; 150 LPC; 152 LPC; 153 mapsofpa.com; 154 BL, London; 155 Getty Images/Hulton Archive; 156-157 BAL/Courtesy of the Council, National Army Museum; 158 Getty Images/Hulton Archives; 159 Scala, Florence/HIP/Public Records Office; 161 AA; 162 Imperial War Museum/Department of Art; 165 The Royal Geographical Society; 166 AI; 167 Scala, Florence/© The Andy Warhol Foundation for the Visual Arts/ARS/DACs; 169 Reproduced by permission of Geographers; 170-171 London's Transport Museum © Transport For London; 173 Map Poster.Com; 174-175 Courtesy of Oxford Cartographers/Peters World map supplied by Oxford Cartographers © Akademische Verlagsanstalt; 176 Science Photo Library/Volker Steger; 177 Science Photo Library/NASA; 178 Science Photo Library; 181 Courtesy of Oxford Cartographers/World Population diagram © ODT INC; 182 NASA/ http://earthobservatory.nasa.gov/Newsroom/ BlueMarble.